African American Chemists: Academia, Industry, and Social Entrepreneurship

ACS SYMPOSIUM SERIES **1381**

African American Chemists: Academia, Industry, and Social Entrepreneurship

Sibrina N. Collins, Editor
Lawrence Technological University, The Marburger STEM Center
Southfield, Michigan, United States

American Chemical Society, Washington, DC

Library of Congress Cataloging-in-Publication Data

Names: Collins, Sibrina Nichelle, 1970- editor.
Title: African American chemists : academia, industry, and social entrepreneurship / Sibrina N. Collins, editor, Lawrence Technological University, The Marburger STEM Center, Southfield, Michigan, United States.
Description: Washington, DC : American Chemical Society, [2021] | Series: Acs symposium series ; 1381 | Includes bibliographical references and index.
Identifiers: LCCN 2021019037 (print) | LCCN 2021019038 (ebook) | ISBN 9780841298385 (hardcover) | ISBN 9780841298378 (ebook other)
Subjects: LCSH: African American chemists.
Classification: LCC QD39.5 .A37 2021 (print) | LCC QD39.5 (ebook) | DDC 540.92/396073 [B]--dc23
LC record available at https://lccn.loc.gov/2021019037
LC ebook record available at https://lccn.loc.gov/2021019038

The paper used in this publication meets the minimum requirements of American National Standard for Information Sciences—Permanence of Paper for Printed Library Materials, ANSI Z39.48n1984.

Foreword

The ACS Symposium Series is an established program that publishes high-quality volumes of thematic manuscripts. For over 40 years, the ACS Symposium Series has been delivering essential research from world leading scientists, including 36 Chemistry Nobel Laureates, to audiences spanning disciplines and applications.

Books are developed from successful symposia sponsored by the ACS or other organizations. Topics span the entirety of chemistry, including applications, basic research, and interdisciplinary reviews.

Before agreeing to publish a book, prospective editors submit a proposal, including a table of contents. The proposal is reviewed for originality, coverage, and interest to the audience. Some manuscripts may be excluded to better focus the book; others may be added to aid comprehensiveness. All chapters are peer reviewed prior to final acceptance or rejection.

As a rule, only original research papers and original review papers are included in the volumes. Verbatim reproductions of previous published papers are not accepted.

ACS Books

Contents

Indexes

Dedication

This book is dedicated in memory of Professor William Moore, who was an amazing chemist, educator, and mentor. Thank you.

Chapter 1

Introduction

Sibrina N. Collins*

Lawrence Technological University, The Marburger STEM Center, Southfield, Michigan 48075, United States

***Email: scollins@ltu.edu.**

Why are the stories featured in this book so important to tell? Why this book, and why now? These biographical narratives highlight the importance of inclusion and equity in the chemical sciences. This chapter provides an overview on the significance of celebrating the intellectual achievements of African American chemists who have contributed significantly to society.

Why This Book, and Why Now?

The year 2020 in America has reached essentially a tipping point with people from diverse backgrounds protesting systemic and institutional racism in society and the tragic deaths of countless African Americans. In response to the social unrest, several organizations, colleges and universities, nonprofits and business leaders have released public statements condemning racism, which is actually a very important first step to make significant changes in our society. However, real policy changes are needed, not just mere words (*1*). Most recently, *Chemical and Engineering News* has published a series of articles focused on the importance of inclusion of African American chemists in the discipline (*2–4*). African Americans represent approximately 13% of the U.S. population, and yet represent a dismal 1.6% of faculty teaching at the top 50 chemistry departments. In addition, a recent report highlights that freshman chemistry courses are an "exit point" for many students of color (*5*). Despite these enormous challenges and obstacles, African American chemists have persisted and made important and significant contributions to our society. Most recently, a new generation of younger chemists have boldly used their agency to highlight the contributions of Black chemists through creative storytelling on social media (*6*).

Over the years, there are have been many efforts to address diversity in chemistry, but the next critical step is to address inclusion and equity within the discipline. There are significant differences between the terms diversity, inclusion and equity. Psychologist Asia McCleary-Gaddy writes (*7*), "Diversity is being invited to the party, but "inclusion" is being asked to dance. If researchers were to extend the said analogy to diversity and equity, it would read; "diversity" is the invitation to the party but "equity" is providing transportation for those who do not have a ride." Specifically, a focus on equity is absolutely needed to address the barriers that exist for so many in STEM (Science,

Technology, Engineering and Mathematics). Thus, this book is about the importance of building a true community within the field of chemistry.

Six Degrees of 'Chemistry' Separation

This book highlights the careers and achievements of a 'small subset' of amazing chemists. There are many other chemists that could have been included in this volume. Previous efforts to highlight the intellectual achievements and important societal impact of chemists included celebrating eminent organic chemist, Dr. Percy L. Julian. His amazing journey in chemistry was celebrated in the film "Forgotten Genius." Dr. Julian faced tremendous obstacles and racism during his career, but he made enormous societal contributions developing therapeutics for treating glaucoma and arthritis (*8*). However, there are many other stories about the careers of chemists that have not been told.

I have many personal connections with the chemists profiled in this book. For example, as a young chemistry major, I discovered chemistry pioneers such Dr. Bettye Washington-Greene, who earned a Ph.D. in chemistry from Wayne State University, where I earned my undergraduate degree. While a graduate student in the Department of Chemistry at The Ohio State University focusing on the photochemistry of transition metal complexes, I also completed a "historical research project" identifying African Americans that were awarded doctoral degrees from the department. One of the OSU chemistry PhD alumni is Dr. Nodie Monroe Washington and to my surprise, as a postdoctoral researcher at Louisiana State University, I worked in the same chemistry laboratory with Dr. Washington's niece!

Over 20 years ago, I recall being thrilled to meet another chemistry pioneer Dr. Richard Evans, who is the first African American to earn a PhD in chemistry from Louisiana State University. Dr. Evans' was a student of Professor Saint Elmo Brady, who is the first African American to earn a PhD in chemistry in the U.S. (1916, University of Illinois). During a phone conversation, I asked Dr. Evans, "You actually took a class taught by Dr. Brady?" Dr. Evans' laughed and responded, "Sibrina, a lot of us took classes from him."

I coauthored "Black Panther, Vibranium and the Periodic Table," which was published in the *Journal of Chemical Education* with Professor LaVetta Appleby in 2018, specifically providing a classroom strategy to engage students with the importance and arrangement of the periodic table of elements (*9*). Furthermore, Marvel Studios' Black Panther celebrates the achievements of women and people of color in the STEM disciplines (*10*).

I collaborated with Dr. Pamela Leggett-Robinson by providing a foreword for her book *Overcoming Barriers for Women of Color in STEM Fields: Emerging Research and Opportunities*, which provides a "STEM road map" to navigate the challenges and successes in academia through engaging narratives from women in STEM (*11*). Finally, Dr. Angela W. Peters was my department chair when I joined the faculty at Claflin University, an HBCU in Orangeburg, SC. Thus, this book *African American Chemists: Academia, Industry and Social Entrepreneurship* is both personal and timely.

STEM Literature: Diversity, Equity, and Inclusion

This book highlights and celebrates biographical narratives of African American chemists, with an emphasis on gender. *African American Chemists: Academia, Industry and Social Entrepreneurship* fits in the genre of book titles such as *Overcoming Barriers for Women of Color in STEM Fields: Emerging Research and Opportunities* (*11*); *African American Women Chemists in the Modern Era* (*12*); *African American Women Chemists* (*13*); the ACS Symposium Series 1256 book, *Diversity in a Scientific Community Volume 2: Perspectives and Exemplary Programs* (*14*); and *Sisters in Science: Conversations with Black Women Scientists on Race, Gender and Their Passion for Science* (*15*). The STEM contributions of African American women have largely been ignored by the scientific community. However, with the success of brilliant books such as *Hidden Figures* (*16*) authored by Margot Lee Shetterly, the media is celebrating the achievements of relatively unknown African American scientists.

In 2005, *Beyond Small Numbers: Voices of African American Chemists* authored by sociologist, Willie Pearson, Jr. brilliantly provides compelling stories of 44 chemists (women and men) who persisted in the field of chemistry (*17*). Also published in 2005, *Successful Women in Chemistry: Corporate America's Contribution to Science* (ACS Symposium Series 907) edited by Amber S. Hinkle and Jody A. Kocsis celebrates the achievements of several women chemists and recognizes the challenges of working in Corporate America. "On a roller-coaster ride, one may experience ups, downs, thrills, excitement, and in some cases—fright; The same can be said for both personal and professional experiences," write Hinkle and Kocsis (*18*).

Inclusive Chemistry Pedagogy

African American Chemists: Academia, Industry and Social Entrepreneurship addresses an unmet need in our society, most notably understanding how African Africans made important contributions to the chemical sciences despite being completely marginalized by society. African American contributions to the STEM disciplines are often overlooked by mainstream media, textbooks and in the classroom.

This book is a resource to empower chemists to address equity and inclusion in the classroom and beyond. The book provides lesson plans that align the stories of these chemists with classroom content in various sub disciplines within chemistry. Thus, chemistry professors will have these important educational tools, namely inclusive chemistry pedagogy, which broadens the image of a chemist in the classroom.

The Black Panther paper mentioned previously is an example of inclusive chemistry pedagogy for general chemistry students. Another example is a lesson plan or learning object (LO) published on the Virtual Inorganic Pedagogical Electronic Resource (VIPEr) website focused on the brilliant inorganic chemistry contributions of Professor Gregory Robinson (*19*). In the appendix of each chapter of this book, the reader will find a lesson plan for classroom instruction. The contributions of chemists of color are often excluded in the chemistry curriculum and textbooks. The lesson plans are to be used by chemistry educators and professors to broaden the image of a chemist in the classroom. The appendix of this chapter includes an inorganic chemistry lesson plan focused on a recent paper published (*20*) in the journal, *The Chemical Educator*, which I coauthored with two of my former students and research collaborators.

Chapter Summary

African American Chemists: Academia, Industry and Social Entrepreneurship highlights chemists that have successfully used their skill sets and educational backgrounds to pursue careers as industrial chemists, professors, department chairs and social entrepreneurs. Furthermore, these chemists have published their research in high-impact journals such as *Analytical Chemistry*, *Heterocyclic Communications*, *Inorganic Chemistry*, the *Journal of Chemical Education*, the *Journal of Physical Chemistry*, the *Journal of Proteome Research*, patents and books. In addition, three chemists have established startup companies addressing equity in STEM. Thus, this book provides the opportunity to broaden the equity conversation and expand the image of a chemist in our diverse and global society.

Appendix: Sample Teaching Chemistry Assignment

Point Group Symmetry and Bonding Theories

Description

In this activity, students will evaluate three structures, namely $[Cu(phenan)_2(NO_3)_2]$ (**1**; phenan = phenanthridine), $[Au(phenan)Cl_3]_2$ (**2**) and $[Cu(dmeobpy)(H_2O)_2][NO_3]_2$ (**3**; dmeobpy = 4,4′-dimethoxy-2,2′-bipyridine). Students will need to obtain the CIF (Crystallographic Information Files) and complete measurements on the three inorganic chemistry structures. The students will determine the point groups and explain the bonding of these complexes using Valence Shell Electron Pair Repulsion Theory (VSEPR), the 18-Electron Rule, and Molecular Orbital (MO) Theory.

Learning Goals

There are various learning goals for this assignment:

1) Determine the point groups of the inorganic complexes.
2) Explain the structures of the inorganic complexes using bonding theories.
3) Successfully obtain CIF files and collect measurement data of the structures.
4) Develop an infographic to explain the fundamentals of a laboratory technique.
5) Critically read a peer-reviewed journal article.

Equipment Needs

Internet connection and computer access.

Implementation Notes

Students will need to download software (e.g. Mercury) for visually evaluating the CIF files for the inorganic complexes. The instructor should provide a tutorial on using the software prior to the assignment. Students will also need to access the journal article and supporting material from the *Chemical Educator* website. The instructor will need to provide a lecture focused on constructing ligand group orbitals for the students to sketch an MO diagram. The article "Construction of Ligand Group Orbitals for Polyatomics and Transition-Metal Complexes Using an Intuitive Symmetry-Based Approach," (Johnson, A.R. *J. Chem. Educ.* **2013**, *90*, pp 56 – 62) is an excellent resource.

Evaluation

Students can be evaluated by submitting responses to the worksheets for this lesson.

Student Handout

The journal article "Effects on Human Non-Small Lung Cancer Cell Growth of Cu(II), Au(III) and Re(I) Complexes Containing N-donor Pyridyl Ligands: A Senior Undergraduate Research Project," (Collins et. al, *The Chemical Educator* **2020**, *25*, 139-145) focuses on the synthesis and characterization of inorganic complexes to serve as potential therapeutics for treating non-small cell lung cancer. The paper evaluates five inorganic complexes, but we will focus on three, namely $[Cu(phenan)_2(NO_3)_2]$ (**1**; phenan = phenanthridine), $[Au(phenan)Cl_3]_2$ (**2**) and $[Cu(dmeobpy)(H_2O)_2][NO_3]_2$ (**3**; dmeobpy = 4,4′-dimethoxy-2,2′-bipyridine). You are required to go to the publisher's webpage and obtain both the article and supplemental material. You are also required to download the CIF files for the three complexes, which you will visually evaluate to obtain measurement data (e.g. bond lengths, angles). Answer the following six questions:

1. Determine the number of *d* electrons for each metal center in the inorganic complexes, **1**, **2**, and **3**.
2. Determine if each inorganic complex obeys the 18-electron rule.
3. What is the electronic geometry (VSEPR) for each molecule?
4. Determine the symmetry point group for each molecule.
5. The cation $[Cu(dmeobpy)(H_2O)_2]^{2+}$ (**3**) has a square planar geometry around the central Cu metal ion and belongs to the C_{2v} point group. This is a situation where the molecular geometry and electronic geometries are not the same. Generate the LGOs (ligand group orbitals) with appropriate symmetry using the hybrid orbitals of the Cu shown in Figure A1.
6. Measure the following bond lengths and angles for the three molecules by completing Table A1. (a) Both $Cu(dmeobpy)(H_2O)_2]^{2+}$(**1**) and $[Au(phenan)Cl_3]$(**2**) have square planar geometries. Are your measurements consistent with the square planar geometry? (b) Are your measurements for $[Cu(phenan)_2(NO_3)_2]$ (**3**) consistent with the O_h geometry? (Be sure to review figure 3 shown in the *Chemical Educator* paper.)

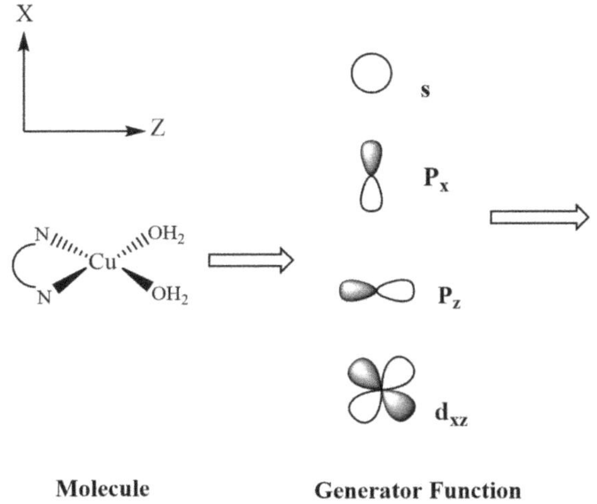

Figure A1. Orbitals (generator functions) of the Cu metal center.

Table A1. Table to Be Completed by the Student

	[Cu(dmeobpy)(H$_2$O)$_2$]$^{2+}$	*[Cu(phenan)$_2$(NO$_3$)$_2$]*	*[Au(phenan)Cl$_3$]*
M-L (Å)			
Cu-N, Cu-O			
Au-N, Au-Cl			
Bond Angles (°)			
N-Cu-N, O-Cu-N			
N-Au-Cl, Cl-Au-Cl			

References

1. Sanford, M. Equity in the Chemical Sciences Requires Actions Not Just Words. *J. Am. Chem. Society*. https://pubs.acs.org/doi/10.1021/jacs.0c06482 (accessed June 19, 2020).
2. Widener, A. The Leaky Pipeline for Black Academic Chemists. *Chem. Eng. News*. https://cen.acs.org/careers/diversity/leaky-pipeline-Black-chemists/98/i22 (accessed June 3, 2020).
3. Howes, L. Essay Criticizing Efforts to Increase Diversity in Organic Synthesis Deleted After Backlash from Chemists. *Chem. Eng. News*. https://cen.acs.org/research-integrity/ethics/Essay-criticizing-efforts-increase-diversity-in-organic-synthesis-deleted-after-backlash-from-chemists/98/web/2020/06 (accessed June 8, 2020).
4. Wang, L.; Satyanarayana, M. Black Chemists Speak Out About Inequity in STEM. *Chemical and Engineering News*. https://cen.acs.org/careers/diversity/Black-chemists-speak-out-about-inequity-in-STEM/98/web/2020/06 (accessed June 19, 2020).
5. Arnaud, C. H. Freshman Chemistry is an Exit Point for Many Underrepresented STEM Students Study Shows. *Chem. Eng. News*. https://cen.acs.org/education/undergraduate-education/Freshman-chemistry-exit-point-underrepresented/98/i23 (accessed Jun 21, 2020).
6. Swiner, D.; Jones, A.; Walker, A. Being #BlackinChem. *Chemistry World*. https://www.chemistryworld.com/opinion/being-blackinchem/4012248.article (accessed Aug 15, 2020).
7. McClearly-Gaddy, A. Be Explicit: Defining the difference between the Office of Diversity & Inclusion and the Office of Diversity & Equity. *Medical Teacher* **2019**, *41* (12), 1443–1444.
8. Wang, L. Percy Julian Film Honored. *Chemical and Engineering News*. https://cen.acs.org/articles/85/web/2007/11/Percy-Julian-Film-Honored.html?ref=search_results (accessed Dec 1, 2020).
9. Collins, S. N.; Appleby, L. Black Panther, Vibranium and the Periodic Table. *Journal of Chemical Education* **2018**, *95* (7), 1243–1244.
10. Collins, S. N. Inorganic Chemistry: Vibranium and Marvel Studios' Black Panther. In *Advances in Teaching Inorganic Chemistry* Volume 1; Jones, R. M., Ed.; ACS Symposium Series 1256; American Chemical Society: Washington, DC., 2020, pp 87 - 95.

11. Robinson-Leggett, P. M.; Villa, B. Overcoming Barriers for Women of Color in STEM: Emerging Research and Opportunities; IGI Global: Hershey, PA, 2020.
12. Brown, J. E. *African American Women Chemists in the Modern Era*; Oxford University Press: New York, NY, 2018.
13. Brown, J. E. *African American Women Chemists*; Oxford University Press: New York, NY, 2012.
14. *Diversity in a Scientific Community Volume 2: Perspectives and Exemplary Programs*; Nelson, D., Cheng, H. N., Eds.; ACS Symposium Series 1256; American Chemical Society: Washington, DC, 2017.
15. Jordan, D. *Sisters in Science: Conversations with Black Women Scientists on Race, Gender, and Their Passion for Science*; Purdue University Press: West Lafayette, Indiana, 2007.
16. Shetterly, M. L. *Hidden Figures*; Harper Collins: New York, NY, 2016.
17. Pearson, W. *Beyond Small Numbers: Voices of African American PhD Chemists*; Elsevier: New York, NY, 2005.
18. *Successful Women in Chemistry: Corporate America's Contribution to Science*; Hinkle, A. S., Kocsis, J. A., Eds.; ACS Symposium Series 907; American Chemical Society: Washington, DC, 2005.
19. Collins, S. N. Gallium Chemistry: To Be or Not to Be a Triple Bond. *VIPEr*. https://www.ionicviper.org/classactivity/gallium-chemistry-be-or-not-be-triple-bond (accessed Oct 16, 2020).
20. Koemm, K. A.; DeBord, M.; Wright, B. D.; Wagers, P. O.; Panzner, M. J.; Stromyer, M.; Youngs, W. J.; Sayandev Chatterjee, S.; Bryan, S. A.; Bonvallet, P. A.; Collins, S. N. Effects on Human Non-Small Lung Cancer Cell growth of Cu(II), Au(III) and Re(I) Complexes Containing N-donor Pyridyl Ligands: A Senior Undergraduate Research Project. *The Chemical Educator* **2020**, 25, 139–145.

Chapter 2

Dr. Bettye Washington Greene: From Tuskegee to the Dow Chemical Company

Sibrina N. Collins*

The Marburger STEM Center, Lawrence Technological University, Southfield, Michigan 48075, United States

***Email: scollins@ltu.edu.**

This chapter focuses on the intellectual achievements and career of Dr. Bettye Washington Greene, who earned a doctoral degree in physical chemistry from Wayne State University in 1965. Dr. Bettye Washington Greene was a research chemist with the Dow Chemical Company in Midland, Michigan for 25 years. The chapter also includes reflections from her daughter, Dr. Willetta Greene-Johnson.

Scientific Achievements and Societal Impact

The scientific achievements and legacy of Dr. Bettye Washington-Greene are truly remarkable. Her doctoral research at Wayne State University in the 1960s provided new knowledge in evaluating light scattering methods to determine the size of particles, which still has important societal implications. Furthermore, her industrial research efforts focused on the chemistry of latex materials, which is also important for meeting societal needs.

Early Years and Education

Bettye Washington Greene was born in Palestine, Texas in March of 1935 to her parents George Washington and Kian Criss. She attended Our Lady of Mercy Elementary School, James E. Guinn Junior High School, and I.M. Terrell High School in Fort Worth, Texas (*1*). She was actually encouraged to pursue chemistry by a high school teacher. During that time in Texas, there were limited career options for African American women; namely one could earn a high school teaching certificate or become a nurse. However, young Bettye had other career goals and enrolled at Tuskegee Institute (now known as Tuskegee University) in Alabama to purse a B.S. degree in chemistry, which she received in May 1955. A few months after earning her B.S. degree in chemistry from Tuskegee, she married William M. Greene (Figure 1) on July 18, 1955. Three children were born from this union, Willetta, Victor and Lisa Kianne.

Figure 1. William M. Greene and Dr. Bettye W. Greene in 1969. (Photo courtesy of Dr. Willetta Greene-Johnson).

"My mother loved Ella Fitzgerald, Sarah Vaughn and Aretha Franklin too," said Willetta Greene-Johnson (2), daughter of physical chemist, Dr. Bettye Washington-Greene. Dr. Willetta Greene-Johnson, who earned a PhD in theoretical physics (3) from the University of Chicago in 1988 and is also a successful songwriter reflects fondly on her parents while she was growing up in Detroit and Midland, Michigan. Her father, William Miller Greene, who was an engineer and former captain in the U.S. Airforce absolutely loved James Brown. Victor and Lisa Kianne chose different successful career pathways. Specifically, Victor worked and retired from the construction industry; Lisa Kianne earned a degree in psychology and works in the service and sales industry.

Music was an important staple in their household on Saturday mornings and her parents would frequently take their children to attend musical concerts. Dr. Greene-Johnson also believes that her brief time living in Detroit (3 years), while her mother pursued her PhD in chemistry at Wayne State University in the 1960s had a tremendous impact on her own musical career as a songwriter. "I was living in Detroit, when a new group called the Supremes was coming up," said Greene-Johnson fondly. Since both of her parents pursued careers in the STEM (Science, Technology, Engineering and Mathematics) fields, they were her first role models. "I lived in a science household and I was not afraid of math."

Graduate Study and Wayne State University

Bettye Washington-Greene made the decision to further her education and earn a doctoral degree, thus she applied for graduate study at both Wayne State University (WSU) and the Massachusetts Institute of Technology (MIT). She was accepted to both graduate programs, but MIT did not offer her financial support (2). Therefore, she began graduate studies at WSU in Detroit, Michigan and was a graduate teaching assistant (1960-1962) and a research associate (1962-1965) in the department (*1*).

The Department of Chemistry at Wayne State University was nationally known as one of the premier research departments in the chemical sciences (*4*). According to historical research from the late Dr. James M. Jay, who was a professor of biology at WSU for many years, Wayne State University was the leading producer of African Americans earning doctoral degrees in chemistry with 13 PhD chemistry graduates (*5*). The University of Chicago (12 graduates) and The Ohio State University (8 graduates) were also leading institutions.

Dr. Greene-Johnson stated that her mother received a world class education at WSU, working directly with Professor Wilfried Heller for her graduate studies. In the acknowledgements of her dissertation (*1*), Washington-Greene writes, "The author wishes to express her sincere appreciation to Dr. Wilfried Heller for his patience, understanding, and invaluable guidance throughout this investigation and to the Office of Naval Research for financial support of the research herein." Thus, effective mentoring and financial support were critical keys for success for Bettye to earn her doctoral degree (*6, 7*).

Born in Bad-Durkheim, Germany in 1903, Heller earned his doctoral degree in 1931 from the University of Berlin (*8*). He held various appointments at the University of Paris, University of Minnesota and the University of Chicago, before joining the faculty in the Department of Chemistry at WSU in 1946. Besides training Dr. Bettye Washington-Greene, Heller trained 25 PhD students in his laboratory focused on colloid and polymer chemistry. Heller retired from Wayne State University in 1974.

Physical Chemistry Doctoral Research

In 1965, Washington-Greene defended her doctoral thesis in physical chemistry entitled "Determination of Particle Size Distributions in Emulsions By Light Scattering (*1*)." Her doctoral research is important today because it helped to establish the necessary techniques and analyses to accurately measure colloidal particle size for a variety of applications including pharmaceuticals, catalysts and cosmetics (*9*). Light scattering is an established technique used to determine the diameter of particles in colloidal suspensions and solutions. In theory, particle diameter is determined from the scattering pattern detected after directing incident light at a solution containing colloidal particles. The ratio of incident light to scattered light provides structural information about the colloidal particles, including particle diameter. The technique works by collecting the average scattering intensity detected after incident light is directed at the sample at different scattering angles, which correspond to different changes in momentum (q) of scattered light (*10, 11*). The momentum, q is defined as follows:

$$q = 4\pi n/\lambda\_0 \ \sin \theta/2$$

where, q is dependent on the refractive index (n) of the solvent, the wavelength of incident light in vacuum (λ_0), and the angle (θ) of the sample positioned relative to the scattering beam.

Washington-Greene's doctoral efforts focused on analyzing the size distribution of 1,6-dimethylnaphthalene (DMN) (Figure 2) colloidal particles in solution. Her data was collected and analyzed at that time using an IBM 7070 Electric Computer (*1*).

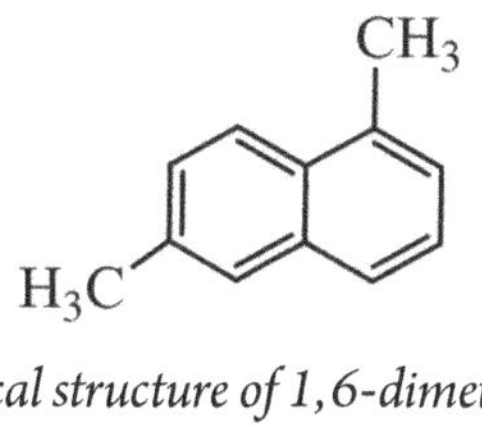

Figure 2. Chemical structure of 1,6-dimethylnaphthalene.

African American Women and Doctoral Degrees in Chemistry

Earning a PhD in chemistry was a remarkable achievement for Dr. Washington-Greene during this time period because very few African American women had earned doctoral degrees in chemistry. Nearly two decades earlier, Dr. Marie Maynard Daly became the first African American woman (*12, 13*) to earn a PhD in chemistry (1948, Columbia University). A few African American women following Daly's achievements were Gladys Williams Royal, who in 1954, earned a PhD from the Department of Agricultural Biochemistry at The Ohio State University (*14*); Katheryn Emanuel Lawson (*15–17*), who earned a doctoral degree in chemistry from the University of New Mexico in 1957; and Reatha Clark King, who earned a PhD in Chemistry from The University of Chicago (*18*) in 1963. (It is particularly interesting to note that Lawson's research efforts in industry were important for validating Crystal Field Theory, which is an electrostatic model focused on the energy splitting of the *d* orbitals in transition metal ions. Lawson also published (*17*) the book *Infrared Absorption of Inorganic Substances* in 1961.)

Additional African American women who earned PhDs in chemistry after Washington-Greene include Esther A.H. Hopkins (*19*) from Yale University in 1967; Gloria Long Anderson (*20*) from the University of Chicago in 1968; Rubye Prigmore Torrey (*21*) from Syracuse University in 1969; Mildred Smalley (*22*) from Louisiana State University in 1972; and Margaret E.M. Tolbert (*23*) from Brown University in 1974.

Industrial Research: Dow Chemical Company

Washington-Greene co-authored a peer-reviewed paper (*24*) with Heller and Mien-Hsiung Chen in the *Journal of Colloid and Interface Science* in 1966, which is critically important because publications in high-impact journals is essentially the "currency of science." After earning her doctoral degree, Washington-Greene was hired as a research chemist in the Edgar C. Britton Research Laboratory at the Dow Chemical Company in Midland, Michigan focusing on latex chemistry (*25, 26*). Dr. Washington-Greene was promoted to senior research chemist in 1970 and three years later, she joined the Designed Polymers Research Division, where she focused on polymer chemistry. Regarding her career at Dow, Washington-Greene did have some challenges. "She went up the ranks, but she was frustrated towards the end of her career," says Greene-Johnson. She was training other employees with less credentials who were being promoted (*2*).

Washington-Greene received several patents and publications focused on latex chemistry research during her career with the Dow Chemical Company (*27–39*). She presented her research findings, which explained the interaction between styrene/butadiene latexes and cellulose fibers, at professional research meetings including the Fourth International Conference on Surface and Colloid Science (*32*) at Hebrew University, Jerusalem, Israel on July 7, 1981.

Prior to Washington-Greene joining the Dow Chemical Company, Dr. Linneaus Dorman was hired as a research chemist in 1961—the first African American chemist to work for Dow in Midland, Michigan (*40*). Dorman, a native of Orangeburg, SC earned his undergraduate chemistry degree from Bradley University (Peoria, IL) in 1956 and was awarded a PhD in organic chemistry from Indiana University (*41*) in 1961.

Dr. Edward Flagg joined the Dow Chemical Company in 1964 as a research chemist (*42*). Flagg, a native of Memphis, TN earned his B.S. degree from Harvard University in 1956. He later earned a M.S. degree from Purdue University in 1959 and a doctoral degree in chemistry from St. Louis

University (*43*) in 1964. Dorman and Flagg also had several peer-reviewed publications and patents (*44–51*).

There were very few African Americans living in Midland, Michigan during this time. In fact, the Greene family was essentially the third African American family to live in Midland, Michigan out of 30,000 residents (*2*).The Greene family often visited with the Dorman family in Midland, MI. Dr. Washington-Greene would later become good friends with Dr. Betty Jones, who is an educator and historian of African American history and culture in Midland, Michigan (*52, 53*). She has specifically conducted research focused on African Americans employed at both the Dow Chemical Company and Dow Corning.

Retirement from Corporate America

Balancing a career as a chemist and wife and mother can certainly be a significant challenge. She often brought work home with her while raising three children. "Early on we learned to clean our rooms and shared light chores like folding clothes and taking out the trash," said Greene-Johnson (*2*). Specifically, Victor and Willetta would clean the kitchen after dinner to help their mother around the house.

However, when her daughter Willetta decided to pursue a career in a STEM field, Dr. Washington-Greene provided solid career advice. "My mother did encourage me. Clearly having a mother as a role model was good," said Greene-Johnson (*2*). In fact, her mother specifically wanted her to apply to Harvard, Stanford, and the University of Michigan. Greene-Johnson earned her undergraduate degree in physics from Stanford in 1979.

Dr. Washington-Greene retired from the Dow Chemical Company in 1990 after 25 years of scientific work in Midland, Michigan. "Upon retirement, Mom and Dad took to the road. They loved visiting their friends and relatives and especially their grandchildren," said Greene-Johnson. "They also enjoyed grilling. My mother grew beautiful roses at just the right pH. Dad's garden provided greens and tomatoes for the neighborhood."

She passed away on June 16, 1995 at the age of 60. After her mother's passing, her daughter wrote a song entitled "Saved," which was featured on the Grammy-winning Brooklyn Tabernacle CD in 2004. "It was written shortly after my mother passed," said Greene-Johnson. Specifically, her daughter spent time reflecting on the true meaning of life. "What is important in life? What's valuable to you? What will I do with the rest of my life? I have a wonderful and supportive family. I am saved."

Chapter Summary

Dr. Bettye Washington-Greene's research efforts continue to make an impact in the chemistry community. Specifically, high-impact papers have recently cited her research in journals such as the *Chemistry of Materials* (*54*) and *Langmuir* (*55*). A SciFinder and Google Scholar analysis of her work indicates well over 100 citations in the literature, which emphasizes the impact of her scientific scholarship. Dr. Bettye Washington-Greene was indeed one of the first African American women to earn a PhD in chemistry in the United States, but she is significant because of her contributions to the field of chemistry. A lesson plan which highlights the significance of Dr. Bettye Washington-Greene's impact in the chemistry community is provided in the appendix of this chapter. This lesson plan can be implemented in the classroom for general chemistry and more advanced chemistry majors. Specifically, for a general chemistry course, the lesson plan can be incorporated when discussing chemical formulas and mass percent composition of compounds. Thus, the lesson plan can be aligned with the content of the course.

Acknowledgments

The author is sincerely thankful to Dr. Willetta Greene-Johnson, Dr. Willie Pearson, Dr. Evelynn Hammonds, Dr. Betty Jones, Dr. Christine Chow, Dr. Matthew Allen, Dr. Paul Jaussen, Dr. Vivian Kao, Dr. Julia Kiernan, Professor Martin Schwartz, Alison Stankrauff and Elizabeth Clemens for their guidance and feedback for this contribution.

Appendix: Sample Teaching Chemistry Assignment

Description

This lesson plan highlights the chemistry contributions of Dr. Bettye Washington-Greene, a former research chemist working for the Dow Chemical Company in Midland, Michigan. Washington-Greene joined the Dow Chemical Company in 1965 after earning a PhD in the field of physical chemistry and retired in 1990. Her research efforts at Dow focused on the chemistry of latex. This activity targets general chemistry and upper level chemistry students. This exercise includes conducting a brief literature review, which will demonstrate chemistry knowledge learned in the classroom. The second part of the activity involves developing a creative infographic about a fundamental laboratory technique. Furthermore, this activity provides an opportunity for chemistry instructors to learn about the diverse backgrounds of students enrolled in their courses.

Learning Goals

There are various goals for this assignment:

1. The student will learn about the background of Dr. Bettye Washington-Greene and write a brief summary highlighting what they learned about her career.
2. The student will critically review an article from *Chemical and Engineering News* focused on latex allergies.
3. The student will apply chemistry concepts learned from coursework.
4. The student will create an infographic focused on a laboratory technique.

Equipment Needs

Internet connection and computer access.

Implementation Needs

The students will need access to the chapter focused on the career and achievements of Dr. Bettye Washington-Greene along with an article published in *Chemical and Engineering News* focused on the chemistry of latex.

Evaluation

Students can complete this lesson by submitting responses to the assigned literature assignments and infographics based on a laboratory technique. Assessment can include student presentations focused on the infographic posters through a virtual gallery walk.

Student Handout

Literature Reading Assignment I

Read the chapter about the career of Dr. Bettye Washington Greene in the book *African American Chemists: Academia, Industry and Social Entrepreneurship*. Write a 400-word essay on what you learned about Dr. Bettye Washington Greene using the following discussion questions:

1. Why did Dr. Washington-Greene pursue chemistry as a major? Who encouraged her?
2. What kind of research did she study as a graduate student at Wayne State University? Why was this research important?
3. Are you interested in pursuing a career as a chemist? Why or why not? Have you conducted research in a chemistry lab with a chemistry professor? Briefly describe your experience and research project.

Literature Reading Assignment II

Read the following article before class:
Burks, R. Key Step Found in Latex Allergy, *Chem. Eng. News*, **2007**, *85(3)*, p16.

Dr. Bettye Washington-Greene focused on the chemistry of latex and polymers, while working as a senior research chemist at the Dow Chemical Company in Midland, Michigan. Latex gloves are commonly used in laboratory settings, but some individuals are allergic to latex. Answer the following questions for class discussion on latex chemistry:

1. What is latex? What is MTB?
2. How does MTB cause an allergic reaction?
3. Why is investigating latex allergies important for physicians and scientists?
4. Create a colorful infographic (mini one-page visual poster) that explains the allergic reaction to latex for general audiences.

General Chemistry Concepts: Mass Percent, Polarity

Dr. Bettye Washington Greene utilized the molecule 1,6-dimethylnaphthalene (DMN) in her research as a graduate student. DMN has the following molecular formula $C_{12}H_{12}$ and a molar mass of 156.22 g/mole.

1. Calculate the mass percent of carbon (C) and hydrogen (H) in this molecule. Please show your work.
2. Examine the chemical structure of DMN (Figure A1). Would you expect DMN to be a polar or non-polar molecule? Explain.

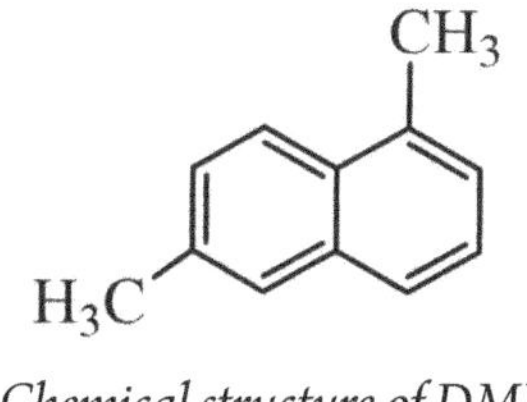

Figure A1. Chemical structure of DMN molecule.

Laboratory Characterization Techniques

1. Examine the chemical structure of 1,6-dimethynaphthalene (DMN). Predict the ^{1}H NMR, FT-IR and MS data you should observe for the DMN molecule.
2. Create an infographic (mini one-page poster) to explain the fundamental concepts of one of the laboratory techniques listed in the above question. Include predicted experimental data for DMN in your infographic.

References

1. Greene, B. W. *Determination of Particle Size Distributions in Emulsions By Light Scattering*. Ph.D. Thesis, Wayne State University, Detroit, MI, 1965.
2. Collins, S. N. *Lawrence Technological University*, Southfield, MI. Personal Communication, 2019.
3. Greene-Johnson. *The Effects of the Exchange Mode Dynamics on Vibrational Phase Relaxation at Surfaces*. Ph.D. Thesis, The University of Chicago, Chicago, IL, 1988.
4. *History-Wayne State Chemistry*. http://chem.wayne.edu/alumni/history.html (accessed June 30, 2020).
5. Jay, J. M. *Negroes in Science: Natural Science Doctorates, 1876-1969*, Balamp: Detroit, MI, 1971.
6. Widener, A. The Leaky Pipeline for Black Academic Chemists. *Chem. Eng. News*. https://cen.acs.org/careers/diversity/leaky-pipeline-Black-chemists/98/i22 (accessed June 3, 2020).
7. Wang, L.; Satyanarayana, M. Black Chemists Speak Out About Inequity in STEM. *Chem. Eng. News*. https://cen.acs.org/careers/diversity/Black-chemists-speak-out-about-inequity-in-STEM/98/web/2020/06 (accessed June 19, 2020).
8. Heller, W. *The Detroit Jewish News Digital Archives*, University of Michigan, July 9, 1982, page 63.
9. Brar, S. K.; Verma, M. Measurement of Nanoparticle by Light-Scattering Methods. *Trends in Analytical Chemistry* **2011**, *30*, 4–17.
10. Bachmann, H.; Brown, A. C.; Clarke, K. C.; Dhada, K. S.; Douglas, A.; Hansen, C. E.; Herman, E.; Hyatt, J. S.; Kodlekere, P.; Meng, Z.; Saxsena, S.; Spears, M. W.; Welsch, N.; Lyon, L. A. Ultrasoft, Highly Deformable Microgels. *Soft Matter* **2015**, *11*, 2018–2028doi:10.1039/c5sm00047e.
11. Dhont, J. K. G. *An Introduction to Dynamics of Colloids*, Elsevier Science: New York, 1996.
12. Collins, S. N. Unsung: Marie M. Daly. *Undark Magazine.*. https://undark.org/article/unsung-marie-maynard-daly-women-stem/ (accessed Sept 27, 2017).
13. Brown, J. E. *African American Women Chemists*, Oxford Press: New York, 2011, pp 34-41.
14. Royal, G. W. *The Influence of Rations Containing Sodium Acetate and Sodium Propionate on the Composition of Tissues from Feeder Lambs*. Ph.D. Thesis, The Ohio State University, Columbus, OH,1954.
15. Lawson, K. *Behavior of Indium at Tracer Concentrations*, Ph.D. Thesis, University of New Mexico, Albuquerque, NM, 1957.
16. Science Couple Finds Success In Albuquerque. *Ebony* **1965**, *20* (8), 67–73. https://books.google.com/books?id=Nd4DAAAAMBAJ&lpg=PA67&pg=PA67#v=onepage&q&f=false (accessed June 30, 2020).
17. Lawson, K. E. *Infrared Absorption of Inorganic Substances*; Reinhold Publishing Corporation: New York, 1961.
18. Clark-King, R. Becoming a Scientist: An Important Career Decision. *SAGE: A Scholarly Journal on Black Women* **1989**, *VI* (2), 47–50.
19. Hopkins, E. A. H. *The History Makers*. https://www.thehistorymakers.org/biography/esther-ah-hopkins (accessed July 2, 2020).
20. Anderson, G. L. *The History Makers*. https://www.thehistorymakers.org/biography/gloria-anderson-42 (accessed July 2, 2020).

21. College Remembers College Professor, Scientist Rubye Torrey. *Syracuse University News*, https://news.syr.edu/blog/2018/10/31/college-remembers-professor-scientist-rubye-torrey-g69/ (accessed July 2, 2020).
22. Smalley, M. R. *Reactions of Hydridoplatinum(ii) Complexes with Acetylenes: A Synthetic and Kinetic Study*. Ph.D. Thesis, Louisiana State University, 1972.
23. Tolbert, M. E. M. *Studies on the regulation of gluconeogenesis in isolated rat hepatic parenchymal cells*. PhD. Thesis, Brown University, 1974.
24. Heller, W.; Cheng, M.-H.; Greene, B. W. Surface Tension Measurements By Means of the Microne Tensiometer. *Journal of Colloid and Interface Science* **1966**, *22* (2), 179–94.
25. Speaking of People. *Ebony* **1973**July, 6.
26. Warren, W. *Black Women Scientists in the United States*, Indiana University Press: Bloomington, 1999, pp 109-110.
27. Makati, A. C.; Lee, D. I.; Greene, B. W.; Iwamasa, Robert, T. *Latex-based Adhesives Prepared by Emulsion Polymerization*. Eur. Pat. Appl. EP 476168 A1 19920325, 1992.
28. Makati, A. C.; Lee, D. I.; Greene, B. W.; Iwamasa, R. T. *Preparation and Uses of Latex-based Adhesives by Emulsion Polymerization*. U.S. Patent 4968740 A 19901106, 1990.
29. Friberg, S. E.; Liang, P.; Liang, Y. C.; Greene, B.; Van Gilder, R. *Colloids and Surfaces* **1986**, *19* (2-3), 249–53.
30. Greene, B. W.; Chen, S. L. *Stable Latexes Having Phosphorus-Containing Surface Groups*. PCT Int. Appl. WO 8502857 A1 19850704, 1985.
31. Greene, B. W.; Chen, S. L. *Stable Latexes Containing Phosphorus Surface Groups*. U.S. Patent US 4506057 A 19850319, 1985.
32. Greene, B. W. Electrokinetic Study of the Interaction Between Styrene-Butadiene Latexes and Cellulose Fibers. *Industrial & Engineering Chemistry Product Research and Development.* **1982**, *21* (4), 592–8.
33. Greene, B. W.; Nelson, A. R.; Keskey, W. H. Redispersible Styrene/Butadiene Latexes. 1. Preparation and Properties. *Journal of Physical Chemistry* **1980**, *84* (12), 1615–20.
34. Greene, B. W. Quantitative Determination of Surface Carboxyl Groups in Vinyl Acid Modified Styrene/Butadiene Copolymer Latexes. II. Latexes Prepared with Methacrylic Acid. *Journal of Colloid and Interface Science* **1973**, *43* (2), 462–72.
35. Greene, B. W. Quantitative Determination of Surface Carboxyl Groups in Vinyl Acid Modified Styrene/Butadiene Copolymer Latexes. I. Latexes Prepared with Acrylic Acid. *Journal of Colloid and Interface Science* **1973**, *43* (2), 449–61.
36. Greene, B. W. Effect of Added Salt on the Adsorbability of a Synthetic Polyelectrolyte. *Journal of Colloid and Interface Science* **1971**, *37* (1), 144–53.
37. Greene, B. W.; Saunders, F. L. In situ Polymerization of Surface-active agents on Latex Particles. III. Electrolyte Stability of Styrene-Butadiene Laxtexes. *Journal of Colloid and Interface Science* **1970**, *33* (3), 393–404.
38. Greene, B. W.; Sheetz, D. P. In Situ Polymerization of Surface-Active Agents on Latex Particles. II. The Mechanical Stability of Styrene/Butadiene Latexes. *Journal of Colloid and Interface Science* **1970**, *32* (1), 96–100.
39. Greene, B. W.; Sheetz, D. P.; Filer, T. D. In Situ Polymerization of Surface-Active Agents on Latex Particles. I. Preparation and Characterization of Styrene/Butadiene Latexes. *Journal of Colloid and Interface Science* **1970**, *32* (1), 90–5.
40. Newman, C. C. Dorman as HistoryMaker: It All Started with a Chemistry Set. *The Daily News*. https://www.ourmidland.com/news/article/Dorman-as-HistoryMaker-It-all-started-with-a-6933041.php (accessed July 7, 2020).

41. Dorman, L. C. *Synthesis of Heterocyclic Nitrogen Compounds of the Tetrahydriimdazole and Hexahydropyrimidine series-reduction of aromatic ketones with alloys of Magnesium*. Ph.D. Thesis, Indiana University, Bloomington, IN, 1961.
42. Flagg, E. *Midland Daily News*, https://www.ourmidland.com/news/article/Dr-Edward-Earl-Flagg-6998938.php (accessed July 9, 2020).
43. Flagg, E. E. *Transition Metal Complexes of Phosphorus*, Ph.D. Thesis, St. Louis University, Saint Louis, MO, 1964.
44. Schmidt, D. L.; Flagg, E. E. Bis(tetrahydrofuran)-haloalanes. *Inorganic Chemistry* **1967**, *6*, 1262–5.
45. Flagg, E. E.; Schmidt, D. L. Poly[tris(diakylphosphinato)alanes]. Novel tris(phosphinates) of aluminum. *J. Am. Chem. Soc.* **1968**, *90* (15), 4173–4.
46. Ridenour, R. E.; Flagg, E. E. Tin-oxygen-phosphorus compounds. *Journal of Organometallic Chemistry* **1969**, *16* (3), 393–404.
47. Schmidt, D. L; Rose, G. D.; Flagg, E. E. *Photocurable Onium Compound Compositions*. US Patent US 5238747 A 19930824, 1993.
48. Dorman, L. C.; Shiang, W. R.; Meyers, P. A. Purification of γ-Benzyl and γ-Methyl L-Glutamate N-carboxyanhydrides by Rephosgenation. *Synthetic Communications* **1992**, *22*, 3257–62.
49. Meyers, P. A.; Dorman, L. C. *Polymer-Containing Agrochemical Emulsion Concentrates*. US Patents US 4818536A 19890404, 1989.
50. Dorman, L. C.; Meyer, V. E.; Wu, M. L. *Gas-Separation Membranes from Polymers Containing a hydrocarbon backbone and pendant (hydrocarbylamido)alkyl ester moieties*. US Patent US 4695295 A 19870922, 1987.
51. Dorman, L. C.; Meyers, P. A. *Hard Tissue Prosthetics*. Eur. Pat. Appl. 192068 A1 19860827, 1986.
52. *Midland's Jones Among Regional 'Drum Major' Recipients, Midland Daily News*, https://www.ourmidland.com/news/article/Midland-s-Jones-among-regional-Drum-Major-6912727.php (accessed July 8, 2020).
53. Collins, S. N. (2020, July 9) Personal Communication (Jones, B.)
54. Moon, S.; Kim, J. Q.; Kim, B. Q.; Chae, J.; Choi, S. Q. Processable Composites with Extreme Capacities: Toward Designer High Internal Phase Emulsions. *Chemistry of Materials* **2020**, *32* (11), 4838–4854.
55. Morilshi, M.; Kitayama, Y.; Ooya, T.; Takeuchi, T. Amphiphilic Polymerizable Porphyrins Conjugated to a Polyglycerol Dendron Moiety as Functional Surfactants for Multifunctional Polymer Particles. *Langmuir* **2015**, *31* (47), 12903–12910.

Chapter 3

Dr. Nodie Monroe Washington: From Louisiana to the Procter & Gamble Company and Beyond

Nodie Monroe Washington*

The Procter & Gamble Company, Cincinnati, Ohio 45202, United States

***Email: nwashington@woh.rr.com.**

This chapter provides brief highlights in my life that enabled me to become a professional chemist. From childhood nurturing from family and teachers at all levels in my pre-college education to my undergraduate college mentors, my journey has been one filled with successes for which I am eternally grateful. These early successes facilitated my work as a graduate student, a college teacher and a professional chemist and manager at Procter and Gamble, a Fortune 500 company. In addition, I share the importance of establishing technical credibility and lessons learned throughout my career.

The Early Years

I was born and grew up in Shreveport, Louisiana during segregation. My father, Early D. Monroe and mother, Laverne Ridley Monroe, worked extremely hard to provide for their four children. My father worked as a stonecutter for a local company that supplied novel marble and other custom-made stones to local companies and individuals. In my early years, my mother was a homemaker and then provided cleaning and food service outside the home when all her kids were beyond elementary school. My parents were intelligent and valued education, even though they did not graduate from high school. It was common to see my mother at our elementary school talking with teachers about how we were doing in school. My parents' focus on education instilled in us the need to continue our education beyond high school. All my siblings have advanced college degrees. This passion for education has extended to the next generation, as my daughter, Dr. Jenene Washington-Stepter, and my nieces and nephews attended college and many have advanced degrees. I look forward to seeing my grandson, Gabriel Jaxon Stepter, usher the legacy of educational excellence into generations to come.

My interest in science started in high school. My biology teacher, Mr. Leroy Ray, required his students to do a science project. My project, Spirogyra and Cobalt Sixty, was entered into local science fairs. After winning first place at several local and regional competitions, I was selected to attend the National Science Fair in Albuquerque, New Mexico. My project was designed to study the absorption profile of Spirogyra (a green algae) toward chemical solutions using the radioactive isotope cobalt 60 and identify its future use in industry for food and other applications. This work exposed me to science students and scientists from all over the United States (including Dr. Wernher von Braun, renowned NASA Scientist) and helped propel my interest in science. In high school, I was a member of the Science Academy and the Phi Beta Chemistry Clubs and a participant in a National Science Foundation (NSF) Summer Science Institute at Grambling College (now Grambling State University), a Historically Black College and University (HBCU). All these initiatives focused on inspiring students to explore scientific careers. During my last year in high school, I took a chemistry class. I loved chemistry and decided this would be the area I would pursue in college.

After graduating as Booker T. Washington High School valedictorian, I entered Southern University, an HBCU in Baton Rouge, Louisiana. The decision to attend Southern University was due to the encouragement of my junior high and high school teachers and administrators who continuously advocated for me, including but not limited to Alphonse Jackson (junior high principal), Bennett Goodwin Turner (junior high band director), Myrtle Stewart (high school English teacher), Leroy Ray (high school Biology teacher), B.H. Fields (high school Chemistry teacher) and many others.

The Undergraduate Years: Southern University

Several months before I entered my freshman year at Southern University, I participated in an NSF Summer Science Institute in the chemistry department at Southern. Along with other high-achieving students from across Louisiana who would enter their freshman year at Southern University, we were exposed to chemistry courses taught by key members of the chemistry faculty. By the end of the summer, we had completed enough chemistry to satisfy our first semester requirement in chemistry. Hence, when we began our freshmen year, we enrolled in second semester courses as a cohort and continued as a cohort through our matriculation as chemistry majors.

As a chemistry major at Southern, I was exposed to many faculty who were awarded doctoral degrees. The chemistry department chair, Dr. Vandon White, took pride in staffing the department with young PhDs who would interact with students and engage them in scientific research. Along with some members of my cohort group, I performed research with Dr. Wilbur B. Clarke, a well-known organic chemist, who earned his PhD from Indiana University in 1963 (*1*). Conducting research in Dr. Clarke's laboratory taught me the importance of scientific curiosity, the need for systematic methodologies and the requirement of questioning and challenging conclusions.

Because of my academic work in chemistry and other classes at Southern, I was selected to participate in another NSF program at the University of Missouri-Columbia during the summer of my junior year. This work led to my first publication involving tritium reactions with benzene (2). This work coupled with my research in the laboratory of Dr. Clarke enhanced my desire to pursue a career in chemistry.

During my senior year, I decided to attend graduate school. At the recommendation of Dr. Vandon White, I chose Rutgers University. Graduating as Summa Cum Laude with a Bachelor of Science in Chemistry in 1968, I enrolled at Rutgers University to pursue graduate studies in Physical Chemistry. I owe tremendous gratitude to many professors at Southern University who facilitated my success including chemistry professors, specifically Dr. Vandon White, Dr. Wilbur Clarke, Dr. Jack Jefferson and mathematics professor Dr. Dolores Spikes.

Graduate Study in Chemistry

Rutgers University

My tenure at Rutgers was relatively brief. Entering Rutgers in the fall after graduating from Southern, I took the required graduate courses in chemistry. I was successful in all of them. I had even begun qualifying to enter the PhD program by passing a series of examinations required by the chemistry department and selecting a research area and professor. I began research with Dr. Joseph Potenza, a physical chemist who was an assistant professor in the department at the time. His research area was Dynamic Nuclear Polarization (DNP) with applications for Nuclear Magnetic Resonance (NMR). DNP focuses on transferring high spin polarization from the electrons of paramagnetic impurities to the host nuclei by means of microwave irradiation close to electron paramagnetic resonance (EPR) frequency. DNP has applications in a variety of areas including particle physics, NMR sensitivity, spin thermodynamics in solids and more (3).

While my academic work was going very well at Rutgers, I decided to obtain the Master of Science degree in Chemistry in January 1970 and leave the university. A key reason was to marry Willie J. Washington, a graduate student I'd met at the University of Missouri during the summer before entering my senior year at Southern. He was a doctoral student pursuing a degree in Plant Genetics. We were married in June 1970, one week after he received his Ph.D. in Plant Genetics.

Prior to the wedding, I taught General Science at a junior high school in Shreveport, LA, my hometown, from January 1970 - May 1970. This was during the desegregation of the public schools in Shreveport, LA. I was among the first group of African American teachers participating in the desegregation of schools in my hometown. That experience reinforced my desire to positively influence children.

The Ohio State University

I entered the PhD program in Chemistry at Ohio State about eight years after leaving Rutgers University. My husband and I moved to Jackson, Mississippi shortly after our wedding. I taught chemistry at Jackson State University, an HBCU, while he was a biology professor at Tougaloo College, also an HBCU.

After a year, we moved to Wilberforce, Ohio where my husband joined the faculty at Central State University, an HBCU. While I began taking chemistry courses at The Ohio State University, I took time off to serve as a laboratory chemist for the state of North Dakota at North Dakota State University in Fargo because my husband accepted a post-doctoral position for a year to do research in Plant Genetics. I developed chemical procedures and performed chemical analyses for a variety of groups including hospitals, criminal agencies and more. During our time in Fargo, our daughter, Jenene Renaye Washington, was born.

While on leave from Ohio State, I later took a faculty position in Chemistry at Central State University, an HBCU. As I mentored many students and encouraged them to pursue graduate studies, I decided to leave my teaching position and focus full time on completing my PhD work in Physical Chemistry at Ohio State. This also coincided with my daughter's entry into kindergarten.

My research area at Ohio State focused on Electron Paramagnetic Resonance (EPR) under the direction of Professor Roger E. Gerkin. My specific work studied the rare earth trivalent gadolinium in single crystals of rare earth bromates largely, with some work in lanthanum magnesium nitrate. The research involved growing single crystals of gadolinium doped rare earth bromates, exploring their electron paramagnetic profiles and obtaining high spin Hamiltonian (H) parameters of the gadolinium ion in the rare earth bromate and nitrate systems.

$$H = T + V_C + V_Z + V_{SO} + V_{SI} + V_{CF} + V_{SS}$$

Where, T represents the kinetic energy of the electrons; V_C is the coulomb attraction of the nucleus and electrons and repulsion of electrons; V_Z is the Zeeman term; V_{SO} represents the spin-orbit coupling term; V_{SI} is the magnetic interaction between unpaired electrons and the nuclear magnetic moments of the same atom (commonly referred to as a hyperfine interaction); V_{CF} signifies interaction between the paramagnetic ion and the crystal field; and V_{SS} is the magnetic dipole interaction of the electrons (4).

Research in the spectroscopy of rare earths provides insights of the fundamental structure of the solid state by observing their behavior in crystalline structures. Rare earth doped crystals are used in lasers (Light Amplification by Stimulated Emission of Radiation) and masers (Microwave Amplification by Stimulated Emission of Radiation). The gadolinium (III) ion was successfully utilized by Scovil and coworkers as the active material in the first operational solid-state maser (5).

The magnetic properties of gadolinium facilitate its use in radiocontrast agents in magnetic resonance imaging (MRI) to improve the diagnostic accuracy of the MRI scan. For example, it improves the visibility of inflammation, tumors, blood vessels and, for some organs, blood supply (6). My doctoral studies in the Department of Chemistry at The Ohio State University were completed in 1982 with a PhD in Physical Chemistry (4). I then assumed a position as a Research Chemist at The Procter and Gamble Company, located in Cincinnati, Ohio.

Professional Years as a PhD Chemist: The Procter & Gamble Company

Having experienced five years of teaching at the college level, I chose to begin a career in industry as a research chemist. I was recruited by The Procter and Gamble Company (P&G) while completing my doctoral degree at Ohio State. I was intrigued by the large variety of consumer products designed and produced by the company. These products provided real-life examples of chemistry applications.

My first project as a research chemist at P&G was in fabric softeners at the company's basic research center in Cincinnati, Ohio. Focused on fabric softener mechanisms and factors impacting softener deposition and efficacy, I found applications from my research work in magnetic resonance spectroscopy extremely helpful. The work led to improvements in the company's fabric softener product, Downy.

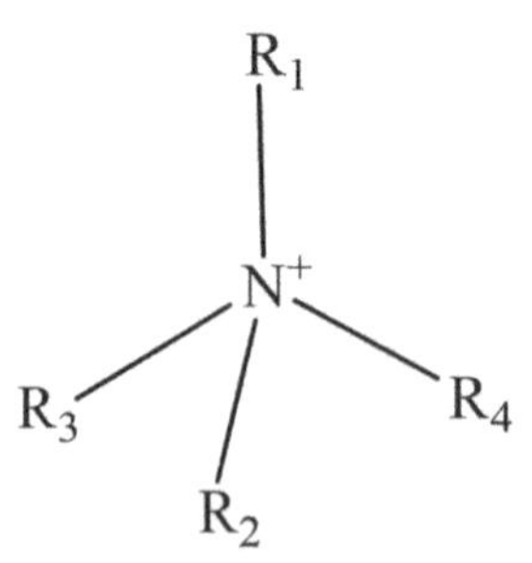

Figure 1. Chemical structure of a quaternary ammonium cation.

The quaternary ammonium cation (Figure 1) was a key ingredient of focus in my mechanistic understanding work. The R groups shown may be the same or different alkyl or aryl groups. Also, the

R groups may be bonding together (*7*). Quaternary ammonium cations are extremely important in various consumer cleaning products (*8*).

My second research project focused on soil release polymers for application in P&G's detergents. Working with a team of scientists, my focus was on the mechanistic understanding of the deposition of soil release polymers onto key fabric types through-the-wash to render easy removal of oily soils during laundering. Block polyesters containing branched hydrophilic capping groups were a key focus of our research efforts (*9, 10*). This work led to key ingredients placed into Liquid Tide and my advancement to Senior Scientist. Subsequently, I moved from the company's basic research facility to their operating division focused on the improvement of Tide and other detergents.

Odor, Bleach, and Fabric Care Technology

Further technology assignments in the company's operating division in Fabric Care led to the enhancement of technologies for color and fabric care, odor control and bleach technology along with my advancement to section head of Odor, Bleach and Fabric Care Technology projects. My odor control project research focused on encapsulated perfume systems for use in detergents. Working with a team of scientists and engineers, studies included investigations of perfume delivery systems such as zeolites; deposition profiles of perfumes and scents in laundry systems; the longevity of perfumes and scents on fabrics and consumer recognitions; perceptions and acceptance of the scent and perform on their fabrics (*11*).

Bleach technology projects focused on enhancement of bleach activator systems such as sodium nonanoyloxybenzenesulfonate (NOBS) (*12–17*). NOBS is an important component of laundry detergents and bleaches (Figure 2). It is a bleach activator for active oxygen sources, allowing formulas containing hydrogen peroxide releasing chemicals (specifically sodium perborate, sodium percarbonate, sodium perphosphate, sodium persulfate, and urea peroxide) to effect bleaching at lower temperatures. My work on NOBS led to product upgrades to Tide with Bleach Detergent. Pernonanoic acid, generated in situ, is the key bleaching species in the detergent with bleach.

Fabric care technologies served as a major focus area for me at P&G. I led teams that developed technologies and provided mechanistic understanding to detergent product ingredients. Technologies such as modified cellulose facilitated minimization of pills and fuzz on fabric and protected the color of fabrics during the wash (*18, 19*). Technologies created and studied, such as modified cellulose, provided improvements to Tide Wearcare detergent and other fabric care initiatives.

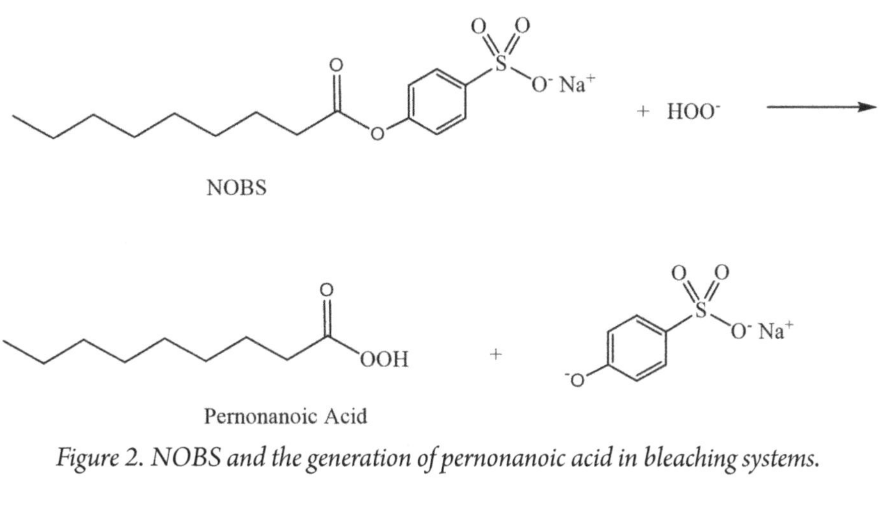

Figure 2. NOBS and the generation of pernonanoic acid in bleaching systems.

Global Textile Chemistry and Fabric Care Research

In order to take fabric care to the next level, I originated a Textile Chemistry and Technology Section at P&G, which ultimately grew into a department with my advancement to Associate Director, a level considered an officer of the company. This enabled a focus on working externally with all facets of textiles (chemists, manufacturers, and organizations all over the world) to identify technologies for use in the company's detergents and fabric care products. Building on fabric care technologies developed for existing laundry products, the Textile organization increased mechanistic understanding of technologies at the fiber and yarn level to allow marketable messaging of products along with the development of relationships with Textile companies, organizations, manufacturers and retailers (*20–23*). I initiated a Textile Care Center at Zhejiang Science and Technology University (ZSTU), located in the heart of China's textile belt, Hangzhou, China (*24*). This allowed the expansion of textile and fabric care research capability to an area focused on apparel production and finishing, specifically facilitating a path for new fabric care products and benefits.

I facilitated the inclusion of P&G into major Textile organizations across the world. Specifically, influencing how textiles are evaluated for fabric care benefits and care label information along policies around textile manufacturing, I played a major role in various textile organizations. These textile organizations include: the American Association of Textile and Colorists, where I served as Chair of the Colorfastness to Washing and Home Laundering Technology Committees and Member of the Executive Committee on Research; The International Organization of Standards, where I served as a US Delegate; The American Society of Testing Materials- Detergent & Textiles Sections, serving as Chair of the Detergents Committee; and The American Oil Chemists Society (*25–28*).

In addition to creating a new Textile Chemistry and Technology area for P&G, I also created standardized testing infrastructure in the detergency area. Developing a centralized evaluation laboratory capable of assessing new and existing detergent and fabric care technologies proved to be a critical component in rapidly developing new technologies. Imaging techniques along with other novel evaluation tools were a key part of the laboratory. To reapply fiber and fabric evaluations techniques across the company, I was a charter member of the Fiber-Fabric Community of Practice for Procter and Gamble. This allowed reapplication of expertise across not only detergents, but hair, tissue/towel, fabric softeners and more.

Mentoring at Procter & Gamble

Mentoring was a key component of my research at P&G. Motivated by important mentors from high school through college, I was determined to use my skills and position in the company to help others. Having been with the company for only six months, I was one of a few Black women in Research and Development (R&D) to sign a letter to the Vice President of R&D encouraging the advancement of Black Women within R&D. Through evaluations by external consultants and discussions with R&D management, Black women began to assume higher levels within R&D. Key to their advancement was mentoring. Similarly, I was a founding member of the Black Technical PhD group within the company. This group advocated for the advancement of Black PhDs up the corporate ladder, along with key business-building assignments. Both the Black Women and Black PhD groups are still active and effective within P&G today.

Navigating Career Challenges

An important component to a successful corporate career for me was understanding the challenges of securing prime research projects and rapidly advancing within the R&D system. Early within my career, this was often a trial-and-error approach. The mentoring groups discussed above facilitated my understanding of the importance of forming a broad network to 1) secure and manage high-profile research projects, 2) increase my face-to-face interactions with senior management and 3) become recognized as a thought leader within the company. My initial network was limited to my first project team in fabric softener research but later expanded to senior scientists and research managers across multiple detergent categories within the same research center. Over time, my work became known to other technical center scientists and their management. Coupled with the mentoring groups discussed above, my interdisciplinary network included managers and researchers across the fabric and home care business unit as well as other business units that championed my work and ultimately my successful advancement up the corporate ladder. Key to forming this network was establishing technical credibility within the company.

Establishing Technical Credibility

As a technical product development manager, I often provided my insights to employees on the importance of building their technical credibility within the company. An example of my discussions on technical credibility with the African American research and development (R&D) scientists and engineers at P&G is provided as follows:

> *As technical individuals we're all familiar with the notion that we want to be known as grounded in scientific principles, communicating our thoughts on technical issues with confidence and authority and in doing so having those who listen to us become convinced or persuaded by our point of view. Further, we want to extend this influence beyond our immediate listeners or audience to those with whom our listeners come in contact by making our listeners users and carriers of our message to others, appropriately attributing the idea to the original source–you or me. We become key contacts and resources to our peers and associates and are often asked to participate in meetings where our opinion will be used as a pivotal selling point to management on a given project. When we find ourselves experiencing these situations, then we have established clear technical credibility in the organization. In most instances, however, the visibility may not be as obvious as I have suggested, but the utilization of the individual as a resource or the individual's name as a symbol for approval are clear signs of an individual having gained technical credibility in an organization. Gaining technical credibility is a goal we should all have in the organization.*
>
> *How Do We Gain Technical Credibility in the Organization?*
>
> 1. *In order to gain technical credibility, you should have confidence in your abilities and a positive mindset that you are technically sharp. We all enter P&G with confidence and a feeling that we are the best; this is based on our internal or external locus of control. In other words, our internal and external experiences have demonstrated to us that we are the best. So our mindsets are definitely channeled in the right direction. The key is to maintain this confidence as you began your interactions with peers, with supervisors and with managers–who view themselves as being the best–based on their calibration points.*

2. *In order to gain technical credibility, you should get calibrated on individuals in the organization who are viewed as Technically Credible. In other words, identify individuals in the organization who are well-regarded technically, and observe how they do their work. Use them as resources to get-on-board in the organization.*
3. *To gain technical credibility, you should focus on basic science and engineering principles in your work and use these as the justification for project planning, execution, and recommendations. This is critical to establishing your own technical credibility in the organization.*
4. *To gain technical credibility, you should propose models for your work and communicate to others and management via this model. We need to quickly establish that we understand the why and how of our project.*
5. *To gain technical credibility, you should have those who have established technical credibility concur with your findings–and both of you communicate this to management. This is a "quick" way to build technical credibility.*
6. *To gain technical credibility, you should build an alliance with technical resources within P&G and outside P&G. Maintain contact with individuals who are viewed as good technical resources via written or oral communications as well as technical meetings.*
7. *To gain technical credibility, you should write technical summaries of your P&G work and if possible, get external publications. Look for opportunities to get patents–write development reports and follow through with patent attorneys to file applications.*
8. *To gain technical credibility, you should provide a leadership role in identifying internal/external consultants who can provide expertise to a given project and utilize them effectively.*

How will you know when you have established technical credibility?

1. *You are sought out by others as a resource.*
2. *Your name is used to management and others to gain agreement to project recommendations.*
3. *Management does not request opinions of others on your work before making decisions on it.*
4. *You are given assignments that require more than execution skills. You assume responsibility for a definable portion of a project or process and work relatively independently to produce significant results that are important to moving the business forward.*
5. *You influence the future of the organization through original concepts/principles and innovations that often lead the organization into major areas of business building opportunities.*

What Can You Do With Technical Credibility?

1. *You will gain a comfort level in the organization which permits you to focus on your project without major concerns around job security.*
2. *You can be more effective in the organization because you are not putting unnecessary energy toward gaining approval from others toward the validation of your work.*
3. *Broader opportunities will be available to you. These can range from: a) a wider range of project selections; b) assignments across multiple functional areas; c) assignments across different sites; and d) project leader roles on multi-functional teams.*

Individuals who worked within my organization advanced largely due to my mentoring and advocacy for them. These individuals spanned all racial, ethnic and gender groups. This work is among my greatest pride when I reflect on my nearly twenty- five years with P&G. I was awarded a

Lifetime Mentoring Award and the P&G Black Leadership Team Lifetime Achievement Award upon my retirement from the company.

Retirement Years

Retirement has allowed me the pleasure of reflecting on my experiences over the last thirty plus years, and to use these learnings to define and navigate my way through what is next for me. I view retirement as being on the mountaintop, namely a place where you are away from the day-to-day details of work life where you can assess the landscape. You are hopefully secure financially and emotionally and you are comfortable with your network (your circle of friends and acquaintances), your network (the work you are currently doing), and net worth.

Currently, everything looks great from my mountaintop (retirement). I have a view of the big picture, not the potholes that cause us to stumble occasionally. Even the stressful times that I endured during my work life are not remembered as problems, but rather teaching moments that made me stronger. From the mountaintop, it's the game-changing moments that first come to mind. I have been asked: what are the things during my working life for which I'm most proud. I respond with the following:

- I'm proud that I made the decision after only six months with the Procter & Gamble Company to become active in the Black Women's efforts to raise the consciousness of P&G that Black Women were not being recognized and rewarded for their achievements. This was a risk. However, it positively changed the landscape of the company and raised the work life of every employee.
- I'm proud I made the decision to work with the Black PhD group at P&G as one of a few Black women to start the Black PhD efforts for the hiring and the advancement of Blacks PhDs within the company. When I joined P&G in 1982, there were less than 20 Black PhDs. The current count is now more than seventy five. Recall, many retire, and others have moved to other companies.
- I am proud that I stepped out of the box to create a new area for the P&G Company in Textiles that would be strategic for new opportunities and innovations for our laundry products.
- I am proud of my mentoring work and my identification of mentors for those who desired them.
- I am proud of being the second African American woman to receive a PhD in Chemistry from The Ohio State University.
- I am proud of being among the first group of African Americans to teach in desegregated schools in my hometown of Shreveport, LA. That experience reinforced my desire to positively influence children.
- I am proud that I made the decision to complete my PhD while teaching at Central State, even with a daughter and husband. I wanted to model the advice I was providing my students.
- I am proud I started my college education at an HBCU, where I was nurtured to envision a life that would be satisfying intellectually, economically, and socially.

Notice I never said anything about developing technologies for Liquid Tide, Tide with Bleach, and other consumer products. I did not mention the projects I worked on or the positions I held. While these are important and valuable, from retirement, they do not seem as game changing the systemic programs and opportunities that transform organizations and individuals; these things are timeless.

So, with these insights, my mission in retirement highlights my involvement with game-changing activities. And, indeed, that is my focus. My activities are now centered around non-profit organizations focused on historically underrepresented individuals and communities. The key is to use the skills I have developed across the more than thirty years of work to develop and implement better systems, processes, and opportunities for under-served communities. My retirement involved church missionary society activities with local homeless shelters, mentoring university students, working with Optimist International via the local Wilberforce-Xenia Optimist Club to serve local youth, and serving on the Xenia Community School Board.

Wilberforce University

Shortly after retirement, I became a volunteer at Wilberforce University, an HBCU in Wilberforce, OH and minutes from my home. Initially I was given the title of Faculty Development Coordinator since I wanted to encourage the faculty to seek out opportunities to expand their capabilities with student learning. I viewed this work as an expansion of my mentoring efforts during my years at P&G. This work led to the publication of three faculty journals highlighting research efforts and teaching techniques by the faculty (*29–31*).

I was asked by the university to assist with their reaffirmation efforts, and ultimately became the leader of this effort (*32*). I became a Special Assistant to the President. Working with faculty, students and the administration along with external groups, we were successful in our reaffirmation with the North Central Association of Colleges and Schools.

In collaboration with the Missionary Society of the Third Episcopal District of the African Methodist Church, I initiated a mentoring program at Wilberforce University. Male and female adult professionals in the greater Wilberforce community were paired with freshmen students with the aim of facilitating the successful transition to the college experience. This effort proved beneficial for both mentee and mentor. During the peak of the program, there were thirty-seven mentors with professional credentials in the community. These professionals mentored 110 students. The program promoted growth for all who participated and provided long-term relationships between mentor and mentee. I was awarded the Doctor of Humane Letters from Wilberforce University for my volunteer work.

Xenia Community Schools

During my retirement years, I became increasingly interested in the capability of our local school system. When a vacancy occurred on the Xenia Community School Board, I applied for the position and was appointed. Focused on student and teacher development, I assisted with identifying the best educational leaders for students attending the Xenia Community Schools. I was a strong advocate for all students having food, not only during school but also on weekends via the Backpack program. I was part of a collaboration between the school board and Xenia City Council to enhance contributions to the Backpack program. I promoted use of Volunteers in the school system by assisting with the school district financing background checks for volunteers. I assisted in the formation of partnerships between community organizations and school personnel to have students participate in Oratorical & Essay contests that not only could improve their communication skills but also result in scholarships. I assisted in the Establishment of the Family Assistance program that allows donations of food and non-food items to assist families in obtaining basic needs.

Optimist International

With the mantra of "Bringing Out The Best In Kids," Optimist International has chapters all over the world focused on the positive development of youth. During retirement, I began leadership roles in the organization in my area. Serving as president of my local club, the Wilberforce-Xenia Optimist Club, and as Lieutenant Governor for the state of Ohio Optimist Club (Ohio District), I assisted in the promotion of youth activities locally and regionally within the state. Chartering college Optimist Clubs at two HBCUs, Wilberforce University and Central State University, both in Wilberforce, Ohio, I was instrumental in exposing many students to the joy of helping elementary and high school students in reading and community service work.

Chapter Summary

In summary, I end this chapter with key lessons I have learned, especially during my professional career. Included in the appendix of this chapter is a lesson plan for educators to use in the classroom focused a project-based learning activity for students. My hope is those who read this chapter may be able to reapply these lessons as they navigate their way through life.

Follow Your Passion

You should identify areas that make you look forward to coming into work every day. This was key to my longevity with P&G. As a chemist, I wanted to use my natural curiosity to identify new areas of opportunity that could positively impact the business. Over time, the Textile area became an exciting white space area for new product ideas and technology. This area became my passion.

Focus on How You Can Impact the Organization

Continuously focus on how you can leverage your strengths to uniquely impact your projects and organization. Identify and leverage what you uniquely bring to the team to rapidly advance the work and project. Draw on your strengths and continuously broaden your strengths and skills to identify opportunities, anticipate issues and solve any pending issues and problems. Open your mind to learning new things that are important foundations to your work. Have confidence in your abilities. Be known for something positive in the organization.

Be Committed to Excellence and Expect Great Things from Yourself and Others

Challenge yourself and others to raise the bar on expectations. Consistently work toward a goal. Constantly ask yourself and others how you can improve or how can you take things to the next level. While it may require significantly more work, the result makes it worthwhile. Expect great things from yourself and secure the resources and skills to achieve excellence. Be committed to the old-fashioned work ethic.

Model the Behavior You Desire

The best way to be a change agent in the organization is to live and breathe the change you desire for the organization.

Be Transparent to All

Many who know me would agree on my transparency. Letting others know your thoughts (even if they are not popular) is key to efficiently moving projects and people to the next level. I am a believer in the statement, 'mean what you say and say what you mean.'

Be a Positive Influence in Diverse Interactions and Collaborations

Collaborate and network with others across functions, affinity groups and more to get feedback on ideas and to learn how others solve problems. Experiencing the diversity of thinking styles can richly influence your strategies/approaches to problem solving.

Be Committed to Helping, Developing, and Empowering Others

Some of my greatest rewards during my tenure at P&G occurred when I helped in the development and growth of others. The pay out on this is huge. These individuals become energized and empowered in their work, moving projects forward to achieve the indicated goal.

Set Priorities

We all have lots we need and want to do. It is important to work with your management to agree and align on priorities.

Be Responsible

Individuals and teams should know that they can count on you to deliver.

Be Persistent

For me this has been key to a successful career. For the research areas I was passionate about, I stayed the course to gain supporters.

Appendix: Sample Teaching Chemistry Assignment

Project-Based Learning: Creating a Detergent with Bleach

Description

Students will develop an understanding of the chemistry of essential components of detergents and how each component functions in a complex aqueous environment to provide consumer-desired benefits such as stain removal, whitening/bleaching, brightening, odor removal/delivery and fabric care. Additional focus on the chemistry of the fabrics in the wash will provide an appreciation of the needed synergy between detergent components and fabric composition.

Learning Goals

There are various goals for this assignment:

Basic Chemistry

Students will identify necessity of detergents and their ingredients:

a. Key structural features and properties of surfactants.
b. Why builders are critical components of detergents.
c. The role of detergent enzymes as they provide cleaning and other benefits.

Bleach System

- Students will identify bleach systems commonly used with bleach-based detergents.
- Students will understand why activated bleach systems provide enhanced through-the-wash color-safe benefits versus other bleach systems.

Perfume/Odor Systems

- Students will identify detergent perfume delivery systems and understand the added value these systems bring to detergents.

Textile Chemistry

- Students will understand the chemical composition of fabrics, clothing, and other textiles in the wash on which the detergents provide benefits, noting especially the need for modified polymers in detergents for enhanced cleaning of synthetic fabrics and clothing.

Equipment Needs

Internet connection and computer access.

Implementation Needs

The students will need access to the chapter focused on the career and achievements of Dr. Nodie Monroe Washington. The instructor may also include a laboratory activity, where students can synthesize a laundry detergent additive (e.g. Mascarenhas, C.M. *J. Chem. Educ.* **2013**, *90*, pp.1231-1234).

Evaluation

Students can be evaluated by completing the worksheets for this lesson.

Student Handout and Group Activity Discussions

Dr. Nodie Monroe Washington research efforts at the Procter & Gamble Company focused on the chemistry of bleach and detergents. Specifically investigating NOBS (nonanoyloxybenzenesulfonate), which is an important component of detergents and bleaching systems.

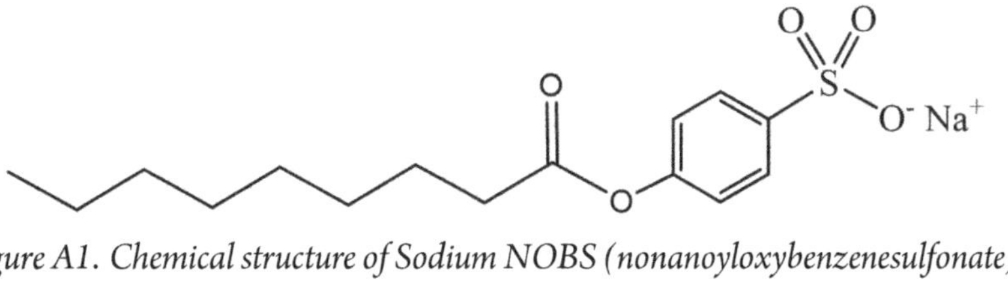

Figure A1. Chemical structure of Sodium NOBS (nonanoyloxybenzenesulfonate).

Gallery Walk and Poster

You will be placed in teams (4 students) and conduct internet research in the four key areas:

1. Basic Chemistry: Identify components of detergents and explain the chemistry (surfactants, enzymes).
2. Bleach Systems: what are the key components?
3. Perfume and Odor Systems: Identify perfume delivery systems and how they work.
4. Textile Chemistry: What are the chemical compositions of fabrics and clothing?

Your team will prepare a creative poster for a gallery walk, which highlights key concepts from each of the four areas.

General Chemistry Concepts

1. The molecular formula and molar mass of NOBS is $C_{15}H_{21}NaO_5S$ and 336.30 g/mole, respectively. Determine the mass percent of oxygen in NOBS.
2. What is the geometry (VSEPR) around the central sulfur atom?
3. What is the predicted hybridization around the carbon atoms in the benzene ring?

Organic Chemistry Concepts

1. You have been hired as a chemist and you need to prepare NOBS in the laboratory. Propose a synthetic strategy for preparing NOBS using nonanoic acid as a starting reactant. Include a mechanism for the formation of the final product.
2. Once you have prepared your NOBS sample, you will need to characterize the product using infrared spectroscopy (IR) and nuclear magnetic resonance (NMR). What evidence would the experimental data provide to confirm the formation of NOBS?

Laboratory Activity (Optional)

You will work in teams to synthesize and characterize a laundry detergent perfume additive.

References

1. Clarke, W. B. *Investigation of McFadyen-Stevens Reaction*, Ph.D. Thesis, Indiana University, Bloomington, IN, 1963.
2. Schroeder, J. W.; Monroe, N. M.; Garland, J. K. Phase Effects on Recoil Tritium Reactions with Benzene. *J. Phys. Chem* **1969**, 73, 1252–1256.
3. Atsarkin, V. A. Dynamic Nuclear Polarization: Yesterday, Today and Tomorrow. *J. Phys.: Conf. Ser.* **2011**, *324*, 012003. https://doi.org/10.1088/1742-6596/324/1/012003 (accessed Sept 16, 2020).
4. Washington, N. M. *X-Band EPR Investigations of Trivalent Gadolinium in Single Crystals*," Ph.D. Thesis, The Ohio State University, Columbus, OH, 1982
5. Scovil, H. E. D.; Feher, G.; Seidel, H. Operation of a Solid State Maser. *Phys. Rev.* **1957**, *105*, 762.
6. Liney, G. *MRI in clinical practice*. Springer: London, 2006; pp. 13, 30.

7. Smulders, E.; Sung, E. Laundry Detergents, 2. Ingredients and Products. In *Ullmann's Encyclopedia of Industrial Chemistry*. Wiley-VCH: Weinheim, 2012. (https://doi.org/10.1002/14356007.o15_o13 (accessed Sept 16, 2020).
8. Howes, L. Rise in Quaternary Ammonium Compounds Observed During Coronavirus Crises. *Chem. Eng. News*. (https://cen.acs.org/environment/pollution/Rise-quaternary-ammonium-compounds-observed/98/i35?utm_source=Twitter&utm_medium=Social&utm_campaign=CEN (accessed Sept 15, 2020).
9. Gosselink, E., Hughes, L., Larrabee, A., Mermelstein, R., Washington. N., *The Procter and Gamble Company, Patent Application, Stable Liquid Detergent Compositions*, Patent US86200584.0, 1986.
10. Gosselink, E., The Procter and Gamble Company, Patent US4,713,194. *Block Polyester And Like Compounds Having a Branched Hydrophilic Capping Groups Useful as Soil Release Agent in Detergent Compositions*. 1987.
11. Pan, R.; You, J. F.; Caravajal, S.; Graves, S.; Mueller, R. *Perfume Delivery System Comprising Zeolites*. U.S. Patent 5,691,303, 1997.
12. Wichman, T.; Getty, E. *Granular Detergent Composition Containing Secondary (2,3) Alkyl Surfactant And A Bleach/Bleach Activator System*. International Application WO 95/05447, 1994.
13. Gosh, C.; Burns, M.; DiGiulio, D.; Getty, E.; Hartshorn. A.; Wiley, A.; Brode, P.; Barnett, B.; Rubingh, D. *Bleaching Compositions Comprising Protease Enzymes*, U.S. Patent 5, 677, 72, 1997.
14. Kuzel, P.; Lieser, T. Bleach systems. *Tenside, Surfactants, Detergents* **1990**, *27* (1), 23–8.
15. Chung, S.; Spadini, G. U.S. Patent 4, 412, 934, 1983.
16. Cahn, A. *Proceedings of the 3rd World Conference on Detergents: Global Perspectives*. The American Oil Chemists Society: Urbana, Illinois, 1994, pp. 64–70.
17. Hirschen, M. *Handbook of Detergents Part C: Analysis*. Marcel Dekker: New York, 2005, pp. 439–470.
18. Leupin, J.; Boyer, S.; Gosselink, E.; Wang, J.; Hunter, K.; Washington, N. *Laundry Detergent Compositions with Cellulosic Based Polymers to Provide Appearance and Integrity Benefits to Fabrics Laundered Therewith*. Patent EP1017726, 1998.
19. Wang, J.; Washington, N.; Hunter, K.; Boyer, S. *Laundry Detergent Compositions with Cellulosic Polymers to Provide Appearance and Integrity Benefits to Fabrics Laundered Therewith*. U.S. Patent 6,833,347B1, 2004.
20. Wang, J.; Washington, N.; Moe, K. *Method for the Use of Hydrophobic Bleaching Systems in Textile Preparation*. U.S. Patent 6830591, 2001.
21. Wang, J.; Washington, N.; Hecht, S.; Miracle, G.; Scheibel, J. *Process for Modifying Textiles Using Ionic Liquids*. U.S. Patent 2006/0090271A1, 2006.
22. Wang, J.; Washington, N. *Textile Mill Applications of Cellulosic Based Polymers to Provide Appearance and Integrity Benefits During Laundering and In-wear*. U.S. Patent 2003/0226212A1, 2003.
23. Wang, J.; Washington, N. Hydrophobic Bleach Systems and Textile Preparation: A Discontinuity in Fabric Care. *AATCC Review* **2002**, *2* (6), 21–24.
24. *Zhejiang Science and Technology University (ZSTU)*, (https://en.wikipedia.org/wiki/Zhejiang_Sci-Tech_University; accessed Sep 18, 2020).

25. *The American Association of Textile Chemists and Colorists.* https://www.aatcc.org/ (accessed Sept 18, 2020).
26. *International Organization for Standardization.* https://www.iso.org/home.html (accessed Sept 18, 2020).
27. *American Society of Testing Materials.* https://www.astm.org/ (accessed Sept 18, 2020).
28. *The American Oil Chemists Society.* https://www.aocs.org/ (accessed Sept 18, 2020).
29. Washington, N., Ed. *Wilberforce University Faculty Journal*, Spring 2008, 1 (1).
30. Washington, N., Ed. *Wilberforce University Faculty Journal*, Fall-Winter 2008-2009, 2 (1).
31. Washington, N., Ed. *Wilberforce University Faculty Journal*, Spring-Summer 2009, 3 (1).
32. Washington, N., Ed. Reaffirmation of Accreditation--An Institutional Lifeline. *Wilberforce University Faculty Journal*, Spring-Summer 2010, 4 (1), 6–7.

Chapter 4

Saint Elmo Brady and Historically Black Colleges and Universities: Preparing Students for PhDs

Richard Arthur Evans†

Alabama A&M University Department of Physics, Chemistry and Mathematics, Huntsville, Alabama 35811-7500, United States

†Deceased

Born in Gulfport, Mississippi in 1938, Dr. Richard Arthur Evans gives a detailed account of his undergraduate experiences at Tougaloo College with his mentor and advisor, Professor Saint Elmo Brady, who is the first African American to earn a PhD in chemistry in the United States. Brady spent his entire career teaching at Historically Black Colleges and Universities (HBCUs). Evans is the first African American to earn a Ph.D. in chemistry from Louisiana State University and served as Department Chair at Alabama A&M University for 26 years. He passed away in 2020.

Undergraduate Years: Tougaloo College

During the summer of 1955 as I made preparation to attend Tougaloo Southern Christian College (now known as Tougaloo College) that fall, I decided to follow my father's suggestion and go into pharmacy. Be mindful that in 1955, at most high schools in the Deep South, there were no counselors to guide students towards careers. This was especially true in the segregated Black high schools where hand-me-down books and supplies from White schools were the rule rather than the exception. Our high school, however, had strong and dedicated teachers, who taught us well in subjects of their specialty. I have no memories of acid-base titrations, periodic tables, weighing exercises or any other physical science experiments because we had no science laboratories, nor did we have courses in chemistry. However, since almost all other Black high school situations were similar, we never considered ourselves as "disadvantaged."

I was glad that my father had suggested pharmacy as a career choice because I had no idea what I wanted to major in or what my career goals should have been. In addition, I had not been given information as to what was required for a career in pharmacy. Upon arriving at Tougaloo in the fall of 1955, I declared pharmacy as my career option, and was immediately told that I would be a chemistry major and that Dr. Saint Elmo Brady would be my advisor. I was also informed that Dr. Brady was not on campus and would confer with prospective advisees upon his arrival from his home in Washington D.C. where he spent his summers catching up on chemistry at the Library of Congress and other readily available and accessible resources.

Mentor, Advisor, and Teacher

Dr. Brady arrived during orientation week, making his first appearance in the dining hall during the lunch period. He was a man of very neat appearance, piercing eyes behind his bifocals, silver-white wavy hair and olive complexion. His pipe and robust laugh all seemed to reinforce his larger-than-life image, which was not at all suggestive of his seventy years. Several of the aspiring chemistry majors descended upon him, as he was about to seat himself at the faculty table, to make ourselves known to him and to make whatever initial impression we could for future consideration.

I can remember telling Dr. Brady that I had not taken a chemistry course, expecting him to begin to enumerate a list of courses to make up for my deficiencies. I was surprised to hear him say, "It didn't make any difference, for if you had, I would have to re-teach you anyway." I then settled down and anticipated taking my first ever chemistry course under the direction of Dr. Saint Elmo Brady.

Departmental Leadership and a Herculean Effort

At Tougaloo, Dr. Brady was a one-person department of chemistry. During my four years as an undergraduate student, I never considered that scenario to be an abnormal. After all, this was a significant improvement over my high school education. Dr. Brady taught a general chemistry class of 40-50 students; a two-semester analytical chemistry class; a two-semester organic chemistry class; a one-semester qualitative organic class; a one-semester biochemistry class and senior research.

To my knowledge, he never used lecture notes. One of his favorite teaching techniques was "chalk and talk" sessions where he would call on students from the class to write and explain assigned topics. General and organic chemistry classes were five semester hours; three one-hour lectures and two three-hour labs per week. Analytical chemistry was four semester hours, with two one-hour lectures, and two three-hour labs. Biochemistry, a three-semester hour course, did not have a laboratory. Qualitative organic analysis, a three-hour course, was mostly laboratory and problem assignments from Shriner, Fuson and Curtin. He managed them all magnificently. There was never an accident or an anxious moment, even when majors would work late at night and sometimes all night.

Dr. Brady utilized the talents of junior and senior laboratory assistants masterfully. We (lab assistants) managed the freshman chemistry laboratories by making up all solutions for first and second semester laboratory courses. Thus, Dr. Brady only had to manage analytical, basic organic and qualitative organic laboratories, still a Herculean task. His teaching load consisted of general chemistry, analytical chemistry, organic chemistry, biochemistry and qualitative organic in alternate semesters and translated to 17 semester hours or 29 contact hours each semester!

Honestly, Dr. Brady taught all of the chemistry courses offered at Tougaloo College between the years 1952-1962. I have concluded that his love of teaching, wealth of experience and dedication to the education of Black youth must have been the major motivating factors for his continuance at Tougaloo after retiring from a long tenure at Fisk University.

Lifelong Commitment to HBCUs

Dr. Brady's teaching career began at Tuskegee Normal and Industrial Institute, now Tuskegee University in Alabama. He joined the staff there as a science teacher in 1908 after completing a bachelor's degree in chemistry at Fisk University, an HBCU in Nashville, TN. The science staff increased from two to six during his tenure and included George Washington Carver. Dr. Brady left Tuskegee in 1912 to matriculate at the University of Illinois at Urbana-Champaign for graduate work in chemistry. He was awarded a master's degree in 1914, and two years later in 1916 became the first African American to earn a doctorate in chemistry (*1*).

When asked about his experiences at Illinois, Dr. Brady would sometimes begin by saying that the year that he started, there were twenty students and one "other (2)." The only colleagues at Illinois that I remember being mentioned by him were polymer chemistry pioneer Carl "Speed" Shipp Marvel and Edward Marion Augustus Chandler. Edward Chandler graduated in 1917, the second African American to earn a doctorate in chemistry (3). Marvel enrolled at the University of Illinois in 1915, and earned his PhD in chemistry in 1920 (*4*, *5*).

Dr. Brady then taught at Howard University, an HBCU in Washington, D.C. where he served as chemistry department head for seven years. He joined the chemistry staff of Fisk University in 1927 and served there as chairperson for twenty-five years. Until the early seventies, only four HBCUs had chemistry departments which were approved by the American Chemical Society (ACS). To his credit, Dr. Brady worked at two of the four HBCUs, namely Howard and Fisk Universities. (The other two were Central State University in Ohio and Morgan State University in Baltimore, Maryland.)

In 1954, the Southern Association of Colleges and Schools (SACS) issued an official order that greatly impacted all HBCUs. Until 1954, majority institutions and historically black institutions were approved using dual standards; one set for majority institutions and one for HBCUs. In 1954, SACS eliminated the dual standards and declared a six-year time schedule for all HBCUs to align with the new standards. It is plausible that Tougaloo College officials must have anticipated the change in SACS standards as early as 1952, when Dr. Brady was asked to join the faculty and serve as chair of the department.

The Impact of Saint Elmo Brady on Chemical Education

It was not until after enrolling at Western Michigan University (WMU) in 1959 to begin graduate work in chemistry that I began to realize the magnitude and extent of the service which Dr. Brady rendered to the education of Black chemists and health professionals. When I arrived on the WMU campus, I was asked two key questions about Tougaloo. First, where is Tougaloo? Secondly, is it accredited? I understood the first question, but the full impact of the second question did not register at the time. The admissions office at Western Michigan gave me conditional admission pending the accreditation status of Tougaloo College. If accreditation had been absent then a list of undergraduate courses in English and mathematics would have been added to my curriculum. Within a few minutes (which seemed like hours), I was given the news that my undergraduate alma mater was indeed accredited to which I breathed a sigh of relief.

While engaged in graduate research at Western Michigan, I wrote to Dr. Brady to explain my research project and to bring him up-to-date on my progress. To my surprise, he answered immediately, offering suggestions for solving my research problem. After leaving WMU, I worked during the 1962-63 school year at Morgan State University, an HBCU in Baltimore, MD.

While teaching at Morgan State in the spring of 1963, we were visited by a classmate who had also been a chemistry major at Tougaloo. We made the decision to go and visit Dr. Brady in Washington, DC. We found Dr. Brady's home address and made a surprise visit to his residence. Dr. Brady received us graciously, brought us up-to-date on the chemistry department at Tougaloo College and engaged us in a two-hour conversation. Unfortunately, that was my last association with Dr. Brady. I discovered through the obituary section in *Chemical and Engineering News* that Dr. Brady died in Washington D.C. on Christmas day, 1966.

Dr. Brady's influence had a direct impact on the aspiration to study chemistry of many graduates of Tougaloo College between 1952 and 1962. Dr. Brady was the kindling fire which was responsible for instilling hopes and dreams within us, and which allowed us to step out and build further on the firm foundation that he established in chemical education.

In February 2019, Professor Saint Elmo Brady was honored by the American Chemical Society (ACS) with a National Historic Chemical Landmark (*6, 7*). Professor Brady taught at four HBCUs and mentored multiple generations of African American chemists, including Dr. Samuel P. Massie, who became the first African American to serve on the chemistry faculty at the U.S. Naval Academy (*8*).

Acknowledgments

Dr. Richard Arthur Evans earned his doctoral degree in chemistry in 1971 from Louisiana State University under the direction of Professor James Traynham. Dr. Richard Evans passed away peacefully at his home surrounded by family on April 15, 2020. The editor is tremendously thankful to the family of Dr. Richard Evans and Dr. Saundra Yancy McGuire.

Editor's Comment

This chapter narrative was authored by Dr. Richard A. Evans. In the acknowledgment section of his dissertation, Dr. Richard Arthur Evans writes, "To Professor St. Elmo Brady, who gave me my initial inspiration and guidance, I express my thanks." Thus, the appendix of this chapter includes a lesson plan focused on the career achievements and legacy of Professor Saint Elmo Brady to be used for classroom instruction. The lesson plan can be targeted to both general chemistry and organic chemistry students. Specifically, for organic chemistry students the lab-based activity can be aligned with class content focused on carboxylic acids.

Appendix: Sample Teaching Chemistry Assignment

The Impact and Legacy of Professor Saint Elmo Brady

Description

In this activity, students will explore the background and career of Professor Saint Elmo Brady, who was the first African American to earn a PhD in chemistry in the United States. Students will screen a brief documentary about his life and career, which includes appearances from members of the Brady family. Students will also review an article published in the ACS Student Member Magazine, *In Chemistry*, which focuses on Professor Brady being honored with a Historic Landmark. Organic chemistry students will work in teams and design their own experiments focused on the synthesis and characterization of carboxylic acids.

Learning Goals

There are various learning goals for this assignment:

1) Explore the background and career of Professor Saint Elmo Brady.
2) Explain the focus of his research project as a chemistry graduate student.
3) Critically read an article focused on Professor Brady's career.
4) Develop a creative poster for a gallery walk focused on the impact of Professor Brady.
5) Design an experiment to focus on the synthesis and characterization of carboxylic acids.

Equipment Needs

Internet connection and computer access.

Implementation Notes

The instructor and students will watch the brief documentary focused on the career of Professor Brady. The students should be required to read the *In Chemistry* Magazine article prior to watching the documentary.

Twenty Whites and One Other Documentary:

https://chemistry.illinois.edu/twenty-whites-one-other-st-elmo-brady-first-african-american-phd-chemistry.

Evaluation

Students will work in teams to prepare a creative poster focused on the career of Professor Saint Elmo Brady. The poster will be presented during a gallery walk. The instructor should provide the students the opportunity to create a rubric for evaluating the posters (content, presentation style, quality).

Student Handout: Literature Discussions, Documentary and Gallery Walk

Literature Discussions

Read the following article before class:

St. Elmo Brady: First African American Chemistry Ph.D., Celebrated Educator Honored with Historic Landmark, *In Chemistry Magazine*. https://inchemistry.acs.org/content/inchemistry/en/acs-and-you/st-elmo-brady.html (accessed Oct 20, 2020).

After reading the article about Professor Saint Elmo Brady, answer the following questions for a classroom discussion. (Answer the following questions prior to the discussion.)

1. Brady's research efforts as a graduate student was a direct result of a disagreement between his research mentor, Professor Clarence Derick and Harvard professor Arthur Michael. Briefly explain this disagreement and include ChemDraw structures with your response.
2. What is the significance of Brady's doctoral research? Briefly explain.

3. After earning his doctoral degree from the University of Illinois at Urbana Champaign, why did he return to teach at Tuskegee?
4. What barriers did Professor Brady face during his career?

Documentary Screening: Twenty Whites and One Other

In class, you will screen the documentary *Twenty Whites and One Other*. After watching the documentary, you will participate in a classroom discussion. What did you learn about Professor Saint Elmo Brady from the documentary? What did you find most interesting and, why?

Gallery Walk

You will work in teams to prepare a creative poster focused on the career and impact of Professor Saint Elmo Brady. A rubric developed by the class will be used for evaluation and grading. Each team poster must include background information (hometown, education) and the significance of his doctoral research with chemical structures and/or reactions. What laboratory techniques would be useful for Professor Brady's research efforts? Your poster should also mention barriers he faced in his career as a chemist. Including a sketch or image of Professor Brady is optional.

Organic Chemistry Laboratory Assignment

You will work in teams to design an experiment focused on the synthesis and characterization of a carboxylic acid. You will prepare an experimental plan with procedures, chemical reagents, reaction conditions and stoichiometric calculations for your overall reaction. The results of your experiment and your proposed reaction mechanism for your product will be presented as an oral presentation to your peers. A rubric developed by the class will be used for evaluation and grading.

References

1. Brady, S. E. *The Scale of Influence of Substituents in Parrafine Monobasic Acids: The Divalent Oxygen Atom*. Thesis, University of Illinois, Urbana, IL, 1916.
2. *Twenty Whites and & One Other*. University of Illinois at Urbana-Champaign. https://chemistry.illinois.edu/twenty-whites-one-other-st-elmo-brady-first-african-american-phd-chemistry (accessed Sept 8, 2020).
3. Chandler, E. M. A. *The Molecular Rearrangement of Carbon Compounds*. Ph.D. Thesis, University of Illinois, Urbana, IL, 1917.
4. Marvel, C. S. *A Study of the Possible Asymmetry of Aliphatic Diazo Compounds*. Thesis, University of Illinois, Urbana, IL, 1920.
5. Leonard, N. J. *Carl Shipp Marvel (1894–1988): A Biographical Memoir*. National Academy of Science: Washington, DC, 1994; pp 233–264.
6. *The American Chemical Society Honors Saint Elmo Brady with a National Historic Chemical Landmark*, https://www.acs.org/content/acs/en/education/whatischemistry/landmarks/st-elmo-brady.html (accessed Jan 10, 2021).
7. *First African American Chemistry Ph.D. honored with landmark during Black History Month*, American Chemical Society, February 4, 2019. https://www.acs.org/content/acs/en/pressroom/newsreleases/2019/february/first-african-american-chemistry-phd-honored-with-landmark-during-black-history-month.html (accessed Jan 10, 2021).
8. Massie, S. P., Jr., *Physics Today*, July 3, 2018. https://physicstoday.scitation.org/do/10.1063/pt.6.6.20180703a/full/ (accessed Jan 10, 2021).

Chapter 5

The Ball Method: A Short Film Celebrating Alice Augusta Ball

Dagmawi Abebe*

Habesha Film Association, 310 S St. Andrews Place, Los Angeles, California 90020, United States

***Email: dagmawiabebe19@gmail.com.**

Why is the story of Alice Augusta Ball so significant to the chemistry community and beyond? This chapter focuses on the revolutionary achievements of chemist Alice Augusta Ball, who became the first woman and African American to earn a master's degree in chemistry from the College of Hawaii in 1915, shattering a tremendous barrier. Ball developed the first effective treatment for Hansen's disease (Leprosy) from the centuries old Chaulmoogra plant. This chapter also includes a discussion about the production of the narrative short film, "The Ball Method," which tells the story of Ball's scientific work on Hansen's disease during her time at Kalihi hospital.

Inspiration for the Short Film: The Ball Method

Imagine a young scientist and filmmaker visiting a university library and being inspired to produce a new short film, while reading a history book on the African American experience in the United States. Libraries serve as critical learning spaces to inspire us to learn about the past and make key historical connections to the present and future. This library visit on that day was my first introduction to Alice Augusta Ball.

I first came across Alice Augusta Ball's (Figure 1) story during the winter of 2017 after reading a book entitled, *The Black West: A Documentary and Pictorial History of the African American Role in the Westward Expansion of the United States* (*1*) at the University of Southern California Doheny Library. I read a paragraph in the book discussing James Ball Sr.'s photographs, which also mentioned that his granddaughter was a revolutionary chemist who was responsible for developing an effective treatment for leprosy (Hansen's disease). Leprosy is a disease that is caused by a bacteria (*Mycobacterium leprae*), which causes significant nerve damage and often leads to disfigurement in patients diagnosed with this terrible disease. Reading this paragraph inspired me to do further research on the life of this exceptional woman.

Figure 1. Alice Augusta Ball. Photo Courtesy of Archives and Manuscripts Collections, University of Hawaii at Mānoa Library. Photo is in the public domain.

Alice's heroic, yet tragic story really resonated with me. First, like young Alice, during my undergraduate years at the University of Virginia, I was an aspiring scientist, specifically majoring in physics. My background as a physicist sparked a curiosity in me to pursue and explore untold scientific stories after I decided to pursue filmmaking during graduate school. Secondly, I was also working on my master's thesis and of similar age as Alice when she passed away at the age of 24. Thus, I decided to focus on Ball's story and submit a proposal to an Alfred P. Sloan Foundation grant program, which provides support for programs and students enrolled in film schools to encourage the emerging generation of filmmakers to address science and technology themes in their work (2). This required me to consultant with the former head of the National Hansen's Disease Center in Louisiana, Dr. David M. Scollard during the writing phase to make sure the scientific ideas and concepts were accurately presented within the grant application. In August 2018, I received the grant to write, direct and produce the very first short film to celebrate the achievements of this amazing young chemist.

Production of the Short Film "The Ball Method"

In 2018, the Alfred P. Sloan Foundation in association with the University of Southern California approved my grant proposal to produce a short film that highlights Ball's achievements in chemistry. After receiving the grant in August of 2018, I took another year to conduct research and narrow the focus of the film to the time period when Alice was already working with Dr. Harry T. Hollmann, an assistant surgeon general at Kalihi Hospital in Hawaii where he was treating patients diagnosed with leprosy. During this research year, I received guidance and critical feedback from my Sloan mentor Thomas Miller and Paul Wermager, former University of Hawaii Science and Technology Librarian, who spent numerous years researching and documenting Alice's life. Furthermore, I traveled to Oahu, Hawaii and the settlement of Kalaupapa on Moloka'i for further research for the short film. These two important visits, especially to Kalaupapa were vital for gathering historical information on how to recreate 1915 Hawaii back in Los Angeles, CA. Surprisingly, most of the settlement of Kalaupapa is preserved just as it was a hundred years ago, with eleven patients still living there and daily tourists passing through, which sustains some of its local economy.

In early August of 2019, after assembling a talented group of cast and crew members, most of whom volunteered their time, I worked closely with producers Mehmet Gungoren and Yeon Jin Lee, on the physical production of the short film. Inspired by Ball's story, the exceptionally talented actress Kiersey Clemons, who could be mistaken for a twin of Alice Augusta Ball, agreed to play her in the film. Collaborating with our production designer Nikki Michelle, we successfully reconstructed an accurate medical ward resembling Kalihi Hospital and Alice Ball's personal living space in the MacDonald's house in 1915. This important detail provides background about daily life inside the hospital and was crucial in aiding to the verisimilitude of the world of the short film. Furthermore, we collaborated with cinematographer, Bash Achkar and we decided to shoot the film in the 4:3 aspect ratio, heightening the feeling of claustrophobia and nostalgia, especially for the scenes at the hospital. This aspect ratio also allowed for seamlessly transitioning from the black and white archive footage provided by the National Hansen's Disease Museum in the opening scenes in of the film.

After six days of shooting and months of post-production, the film premiered at the 2020 Pan African Film Festival. Subsequently, "The Ball Method" was screened at the Sarasota Film Festival; the Bronze Lens Film Festival, where it was nominated for Best Student Film; the African American Film Market; and received the Jury Award for the African American Directors category at the 2020 Directors Guild of America Student Film Awards. The short film is an important example of cinematography, which highlights the scientific contributions from BIPOC (Black, Indigenous, and people of color) communities. The film has been featured on podcasts such as NPR's Science Friday and iHeartRadio's Surprisingly Brilliant. Furthermore, Alice Ball's contributions and the film has been celebrated in publications such as *Massive Science*, *Chemistry World*, and *New Scientist*.

"The Ball Method" as a Film Case Study for Chemistry Majors

Identify a Historical Figure for the Short Film

How can you prepare your chemistry majors to write, produce and direct their own short films? A critical first step in planning to create a short film about a historical figure is taking a deep dive in the literature, then making critical decisions to specifically narrow the focus for the film. This is especially important when making a short film, which is typically 5-20 minutes in length. (I highly recommend for your students to create a 5-minute short film about a historical figure in chemistry.) After researching and learning about Alice Ball's biography, I made the decision to only focus on the time after Alice decided to work on the treatment for leprosy. Selecting this moment in her life allowed me to combine the background world of the numerous leprosy patients suffering at the Kalihi hospital, waiting to be exiled to a remote island and contrasting with Alice Ball's personal life at home and her professional life as a chemistry professor at the College of Hawaii.

Writing a Screenplay for the Short Film

The next step is writing the story in a screenplay format so that everyone involved in the production of the film will have a plan to follow. Paying attention to writing accurate period language is key during this process. It is critical to read old newspaper articles and books to help with writing the screenplay or script. An article published in the *Journal of Chemical Education* provides useful guidelines for YouTube writing and video production, which may be helpful for your students (3). I also recommend for your students to watch other documentaries and narrative films, such as the recent film (4) about the life and career of Dr. Saint Elmo Brady, the first African American to earn a PhD in chemistry in the United States.

Assembling a Cast and Production Crew

Once the writing phase is complete, it is imperative to assemble a team of talented crew members to lead the various productions departments. Your students will be creating 5-minute narrative films, thus the size of their production crews will likely include only 2-3 team members. For our narrative film, I worked with other producers, and a production designer, who focused on recreating the world of the film, including period accurate props and locations. (A production designer may not be needed for the 5-minute documentaries.) We also worked with a cinematographer, who focused on the look of the film, testing various camera lenses, and deciding on a visual language and structure to best communicate the film to an audience. Other department heads such as the costume designer, make-up artist, and art director worked closely together to recreate an accurate depiction of Hawaii in 1915. The common thread between all of these departments is that they are responsible for performing research in their specific arena, sometimes even more than the director as it requires an even narrower focus on a specific subject for the film. Finally, we worked with a chemistry graduate student, who consulted the team about appropriate laboratory equipment, chemistry concepts and experimental procedures which were depicted in the film. Thus, the documentaries can provide your students with an opportunity to discuss important chemistry concepts learned in both the classroom and the laboratory.

Simultaneously, auditions were held to cast the roles written in the film along with the location scouting. Once the film is cast and a filming schedule and locations have been confirmed, you can begin filming the documentary. We filmed "The Ball Method," in six days at four locations. Our chemistry consultant was also vital during this process and worked a great deal with the recreation of the laboratory and classroom settings.

Post-production of the Film

After filming is complete then post-production focuses on carefully editing the footage to tell a compelling story. This process was completed in five months for "The Ball Method." The editing process can be described as writing with images. It is the last chance for the film to be "re-written" and requires a longer time to determine the most effective storytelling method to present to an audience. I decided to begin "The Ball Method" with vintage footage, and so I reached out to representatives at the National Hansen's Disease Museum, who provided me with educational materials that ended up making the final film. Starting the film with archival footage allows for the audience to connect with the time period while providing necessary background information about the world. If your students are going to use archival footage and images, they must receive permission to use the images and provide appropriate acknowledgment for these resources at the end of the film.

Once the picture editing is done, the process of online editing must be completed as well. Online editing consists of color correcting, sound editing and mixing, music scoring, and visual effects. Color correcting refers to the final adjustment and balancing of saturation, brightness, and contrast of the visual images in the film. Sound editing and mixing is a process of "cleaning" the audio track so that the dialogue in the film is prominent when combining sound effects and music. It is imperative to use copyright free existing music or a new music score written by a composer. It is critical that the music is not distracting but enhances the storytelling in the film. Lastly, visual effects are added to enhance a story and fix mistakes that occurred during production. In "The Ball Method," we used visual effects to create a needle being injected into patients in the hospital testing scenes. We also used visual effects to name one of the modern buildings "Department of Chemistry" in the film. These small details helped create an authentic world for the audience. Once post-production is complete, the film is then properly prepared for viewing and screening purposes. Some colleges and universities provide platforms such as YouTube channels for students when then they enroll, which can be used for uploading videos for class projects and other assignments. These online platforms can be used to screen videos in class and virtually to reach wider audiences.

Early Years and Education: Alice Ball

The early years and education of Alice Augusta Ball were brilliantly documented by Wermager and Heltzel (5) in American Chemical Society's *ChemMatters* in 2007. Ball's chemistry career has also been described in other publications as well (6, 7). Thus, my goal of this chapter is to merely highlight some important historical events and background in Ball's very short, but impactful life as a chemist.

Alice Augusta Ball was born in July of 1892 in Seattle, Washington to her parents James Presley and Laura Louise (Howard) Ball (5). She had two older brothers, William and Robert, along with a younger sister named Addie. Her family was considered middle class to upper-middle class, as Ball's father was a newspaper editor, photographer, and a lawyer. The family moved briefly to Hawaii due to the health of her grandfather, but returned to Seattle, WA after his death.

After graduating from high school, young Alice attended the University of Washington in Seattle, where she graduated with two undergraduate degrees: Pharmaceutical Chemistry (1912) and Pharmacy (1914) (5, 6). Furthermore, while attending the University of Washington, Ball along with her research advisor co-authored a paper in the high-impact *Journal of The American Chemical Society* titled, "Benzoylations in Ether Solution (8)," shattering an important first glass ceiling as a female African American scientist. As a result, these accomplishments brought invitations from the University of California-Berkley and the College of Hawaii (now known as the University of Hawaii) for graduate study. Ball would choose the latter and move back to Hawaii to pursue her master's degree in chemistry in 1914.

Graduate Study and the College of Hawaii

During her time in Hawaii, Alice Ball stayed at the YWCA residence, called the "Homestead," and then later moved to the MacDonald's, which was a hotel for women (6). One year later, Ball received her master's degree in chemistry from the College of Hawaii on June 1, 1915, becoming the first woman and African American to attain this achievement (5, 6). Her master's thesis titled, "The Chemical Constituents of the Active Principle of the Ava Root," focused on identifying the active chemical components of the Kava root (9). Ball's laboratory experience would later play a significant role to conduct research to help patients diagnosed with leprosy. Immediately after graduation, Ball accepted an offer to become an instructor at the College of Hawaii's Department of Chemistry, which once again is a significant achievement for an African American woman during this time period.

Research on Hansen's Disease and Kalihi Hospital

Dr. Harry T. Hollmann, an assistant surgeon at Kalihi Hospital in Hawaii as well as the medical doctor at the federally funded Leprosy Investigation Station at Kalaupapa was focused on developing effective treatments for leprosy (5), specifically research efforts focused on Chaulmoogra oil, which has been known for centuries as a possible treatment for this disease. One of the problems with this oil was that it was highly viscous. Upon injection of the oil, many patients would describe seeing this viscous oil as it moved through the skin while also causing a burning sensation (*5*). The active chemical component is Chaulmoogric acid, which is a carboxylic acid (Figure 2).

After coming across Ball's thesis on the Kava plant and because of her expertise in research, Dr. Hollmann asked Ball to join his research lab and develop an injectable solution of Chaulmoogra oil. Hollmann, who published an article in 1922, writes, "I interested Miss Alice Ball, M.S., an instructress in chemistry at the College of Hawaii in the chemical problem of obtaining for me the active agents in the oil of chaulmoogra. After a great amount of experimental work, Miss Ball solved the problem for me by making the ethyl esters of the fatty acids found in the chaulmoogra oil, employing the technique herewith described (*10*)."

Alice worked hard, doing research at night while teaching classes during the day. Her solution involved preparing the ethyl esters of the fatty acids present in the oil. The ethyl ester of an acid is formed using ethanol as the alcohol (Figure 2). Ball's ethyl esters reduced the viscosity of the chemical active compounds in the oil. Therefore, at the age of 24, Alice made a significant scientific breakthrough developing the first viable treatment for leprosy. Thus, another significant milestone achievement.

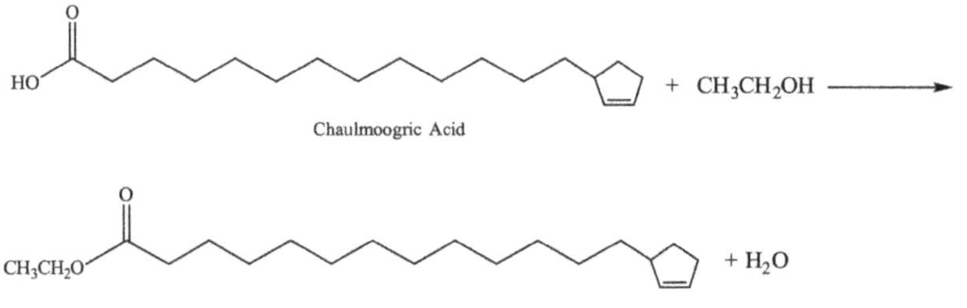

Figure 2. The esterification of chaulmoogric acid.

How Did Alice Augusta Ball Characterize Her Reaction Products?

What type of laboratory equipment and glassware did Ball use to conduct her research? During the early 20th century, chemists used laboratory equipment, which is typically used today, as illustrated in Figure 3. Common glassware used included burets, separatory funnels, reflux condensers and a distillation apparatus. Furthermore, water baths, Bunsen burners, desiccators and Büchner vacuum filtration funnels (*11*) were used during this time period.

Ball's research collaborator, Dr. Harry T. Hollman described in great detail the 'Ball Method' to isolate the active ethyl esters from Chaulmoogra oil (*10*). Initially, Ball used 200 grams of Chaulmoogra oil as starting material, where she isolated crystals of Chaulmoogric acid by vacuum filtration after several recrystallizations from an alcoholic solution. Subsequently, this product was converted to ethyl esters, as shown in Figure 2. In the film, we focused on the importance of recrystallization, which is a critical technique for purifying products from a chemical reaction.

During Ball's era, characterizing products from a chemical reaction was a difficult and tedious process. Chemists primarily used combustion analysis to determine empirical formulas and melting point analysis (*12*). Chemists used other methods as well. In fact, Ball's 1914 paper published in the Journal of the American Chemical Society described using the Volhard method to identify halogens in organic compounds (*8*). Today, chemists use many experimental tools for structural elucidation such as nuclear magnetic resonance (NMR), infrared spectroscopy (IR) and X-ray crystallography. Thus, Ball's accomplishments during the early 20th century were remarkable given the limited experimental tools available to chemists.

Figure 3. "University of Hawaii Research Lab, Gartley Hall" n.d., Office of University Relations Department Collection - Activities - Chemistry, Archives and Manuscripts Collections, University of Hawaíi at Mānoa Library. Photo is in the public domain.

Societal Impact and Controversial Death

One of the tragic aspects of Alice Ball's story is the fact that she was never able to see the full impact of her research on the numerous patients suffering from leprosy. This was because she fell ill and returned home to Seattle, falling in the care of her parents. Her father, James Ball wrote in a letter to the College of Hawaii Librarian Mrs. Elizabeth L. Bryan, Sc.D., "Alice arrived home on the night of the 7th... Dr. Russell has just informed me that Alice is suffering from a severe case of Bronchitis... and it is predicted that in the course of a couple of months that she will be entirely free from all ailments affecting her now (*13*)." Unfortunately, that was far from the reality as Alice Augusta Ball passed away on December of 1916 at the young age of 24 (*14*). The Honolulu Pacific Commercial Advertiser published an article in 1917 that claimed that "While instructing her class in September 1916, Miss Ball suffered from chlorine poisoning (*5*)." This classroom accident is the likely cause of Alice's death as ventilation hoods were not a mandatory requirement for laboratories during this time period.

Therefore, following Ball's death, chemist and president of the College of Hawaii, Dr. Arthur L. Dean continued Alice's work, going as far as having a laboratory at the school manufacture large quantities of the now injectable Chaulmoogra solution to supply the numerous global requests for the new treatment. Many patients were given permission to leave the island of Moloka'i and reunite with their families and loved ones. Sadly, Dr. Dean neglected to cite Alice Ball's work and contribution when publishing his findings in his article, "Preparation of Chaulmoogra Oil Derivatives for the Treatment of Leprosy" and the method was renamed from The Ball Method to The Dean Method (*15*). Dr. Hollmann recognized her significant achievements by highlighting Alice's contributions in his published work (*10*). It took years before she was recognized for her revolutionary scientific achievement (*5–7*).

The Significance of Ball's Legacy

Ball's historical record was kept alive due to the research of Dr. Katheryn Waddell Takara, Mr. Stanley Ali, and Mr. Paul Wermager's persistence and dedication. Dr. Takara was a professor at the

University of Hawaii who started doing research on Black women in Hawaii and discovered Alice Ball's story in 1977. Ali was a retired federal worker who was also doing research on Black women and the catalyst behind the portrait of Alice Ball hanging in the Hamilton Library on campus. In February of 2000, the Governor of Hawaii issued a proclamation declaring February 29th as Alice Ball Day (*5*). In December 2006, the Board of Regents of the University of Hawaii honored Alice's work and memory with its Regent Medal of Distinction (posthumously conferred). In 2008, the University honored Mrs. Ball with a plaque placed at the lone chaulmoogra tree on campus. Most recently, the satellite, ÑuSat 9 also named after Alice Ball was successfully launched into space (*16*).

Chapter Summary

Alice Augusta Ball's story was truly an inspiration to me as a scientist. In the early 20th century, this young African American woman published her undergraduate research in a very prestigious chemistry journal, earned a master's degree in chemistry and became a chemistry professor. Subsequently, she developed the first effective treatment for leprosy (*5–7, 17*) which was a serious public health crisis during her lifetime. This is an enormous scientific achievement!

Alice Ball's method was the most effective treatment for leprosy until the discovery of sulfur-based antibiotics in the 1940s. This provided needed therapeutics for patients and inspired chemists to continue their research efforts for developing more effective pharmaceuticals for treating leprosy. Ball's research was focused on helping patients suffering from leprosy, while simultaneously fighting gender and racial barriers, which unfortunately still exists for women and BIOPIC communities. Alice's story is also important in highlighting the importance of addressing equity and inclusion chemistry.

A lesson plan which highlights the significance of Alice Ball's impact in society is provided in the appendix in this chapter. This lesson plan can be implemented in the classroom for general chemistry and organic chemistry students. Specifically, organic chemistry students will design their own experiments to synthesize and characterize an ethyl ester using various experimental techniques. Furthermore, students will create their own documentaries following the film case study described in this chapter to highlight the significance of an unknown chemist.

Appendix: Sample Teaching Chemistry Assignment

Description

Filmmaker Dagmawi Abebe describes in his chapter the process for creating the very first short film, "The Ball Method" focused on the significant achievements of Alice Augusta Ball. This activity includes a literature reading assignment and students working in teams to design their own experiments to synthesize and characterize an ethyl ester. Furthermore, students will create a 5-minute documentary, highlighting the achievements of an African American chemist using their cell phones as a classroom project.

Learning Goals

1. The student will learn about the background of chemist Alice Augusta Ball and write a brief summary highlighting what they learned about her career.
2. The student will produce a 5-minute documentary about an unknown African American chemist such as Dr. Bettye Washington-Greene and Dr. Dr. Richard Evans.

3. Organic chemistry students will design their own experiments to synthesize and characterize an ethyl ester using experimental techniques learned in the laboratory.
4. Organic chemistry students will develop a creative infographic, which highlights the synthesis and characterization of the ethyl ester research project.

Equipment Needs

Internet connection and computer access.

Evaluation

Assessments can include evaluating student essays about Ball's life and career. The five-minute documentaries on another chemist can be evaluated using a rubric developed by the students. Assessments of organic chemistry students will focus on the content and oral presentation of their creative infographics based on their experiments to synthesize and characterize an ethyl ester.

Implementation

The students will need access to the chapter focusing on the career and achievements of Alice Augusta Ball authored by Dagmawi Abebe.

Student Handout

Literature Reading Assignment (General Chemistry)

Read the chapter about the career of Alice Augusta Ball in the book *African American Chemists: Academia, Industry and Social Entrepreneurship*. Write a 400-essay on what you learned about Alice Augusta Ball using the following discussion questions:

1. Who encouraged Ball to develop a treatment for leprosy? Why was she asked to conduct research on this project?
2. Why is Alice Augusta Ball's research so important to society?
3. What did you find most interesting about Ball's story and why?

Creating Your Own 5-Minute Documentary (General Chemistry and Organic Chemistry)

If you could create your own 5-min documentary about the career and impact of unknown chemist, who would you select, and why? You can use your cell phone to record your own short film to tell a compelling story about a chemist. Your short film will be screened by your instructor and peers.

The first step to creating your own documentary is to watch other documentaries, which will help you to create your own. It is recommended that you watch the 5-min documentary about the career of Dr. Saint Elmo Brady, who is the first African American to earn a PhD in chemistry in the United States. You can access the film at the following link:

Twenty Whites and One Other Documentary

https://chemistry.illinois.edu/twenty-whites-one-other-st-elmo-brady-first-african-american-phd-chemistry

In order to create your documentary, you will need to develop a plan as follows:

1. Identify a historical figure for your documentary.
2. Develop a script or screenplay.
3. Assemble your production team.
4. Film your documentary with your cell phone.
5. Post-production and editing.

Organic Chemistry Infographic and Designing Your Own Experiment

You will work in teams to design an experiment to prepare an ethyl ester. You will need to review the scientific literature and develop a detailed experimental plan for your research project. The experimental plan must be modified from the published procedure. For example, your team can try different solvents or starting materials for the research project. The ethyl ester must be characterized using ^{1}H NMR (Nuclear Magnetic Resonance), Infrared (IR) and Ultraviolet-visible (UV-Vis) spectroscopy.

You will develop a creative infographic that explains the synthesis and characterization of the ethyl ester isolated from your research. Your infographic could be similar to a one-page comic book! Your infographic must include the overall reaction mechanism, reaction conditions, experimental results, and references cited.

References

1. Katz, W. *The Black West: A Documentary and Pictoral History of the African American Role in the Westward Expansion of the United States*. Broadway, 2005.
2. *Film. Alfred P. Sloan Foundation*, sloan.org/programs/public-understanding/film, (accessed Dec 31, 2020.)
3. Franz, A. K. Organic Chemistry YouTube Writing Assignment for Large Lecture Classes. *Journal of Chemical Education* **2012**, *89* (4), 497–501.
4. *Twenty Whites and One Other*. University of Illinois at Urbana-Champaign, https://chemistry.illinois.edu/twenty-whites-one-other-st-elmo-brady-first-african-american-phd-chemistry (accessed Apr 1, 2021).
5. Wermager, P. and Heltzel, C. Alice A. Augusta Ball Young Chemist Gave Hope to Millions. *ChemMatters*, American Chemical Society: Washington, DC, Feb. 2007, pp. 16–19.
6. Brown, J. *African American Women Chemists*. Oxford-United Kingdom, United Kingdom, Oxford University Press, 2011.
7. Jackson, M.. *Alice Augusta Ball (1892-1916)*. https://www.blackpast.org/african-american-history/ball-alice-augusta-1892-1916/ (accessed Feb 25, 2020).
8. Dehn, W.; Ball, A. A. Benzoylations in Ether Solution. *Journal of the American Chemical Society* **1914**, *36* (10), 2091–21001DOI:10.1021/ja02187a015.
9. Ball, A. A. *The Chemical Constituents of the Active Principle of the Ava Root*. Thesis, The College of Hawaii, 1915.
10. Hollmann, H. T. The Fatty Acids of Chaulmoogra Oil in the Treatment of Leprosy and Other Diseases. *Archives of Dermatology and Syphilology* **1992**, 5 (1)doi:10.1001/archderm.1922.02350260097010.

11. Jensen, W. B. The Origins of the Hirsch and Büchner Vacuum Filtration Funnels. *Journal of Chemical Education* **2006**, *83* (9), 1283.
12. Seeman, J. I. Review of the book Classical Methods in Structure Elucidation of Natural Products by Reinhard W. Hoffmann. *Bulletin for the History of Chemistry* **2019**, *44* (1), 68–73.
13. Ball, J. *Letter to Elizabeth L. Bryan*. 11 Apr. 1916. Personal collection of Paul Wermager.
14. Washington, State of. 1917 (filed 2 Jan). "*Certificate of Death: Alice A. Ball*." Board of Health, Bureau of Vital Statistics, Reg. Dist. No. 96, File No. 2585.
15. Dean, A. L.; Wrenshall, R. Preparation of Chaulmoogra Oil Derivatives for the Treatment of Leprosy. *Public Health Reports (1896-1970)* **1922**, *37* (23), 1395doi:10.2307/4576415.
16. Ten Satellogic Earth-Imaging Satellites Successfully Launched – Spaceflight Now. *Spaceflight Now*, https://spaceflightnow.com/2020/11/06/ten-satellogic-earth-imaging-satellites-successfully-launched/ (accessed Jan 8, 2021).
17. How the Woman Who Found a Leprosy Treatment Was Almost Lost to History. *National Geographic News*, 28 Feb. 2018, https://www.nationalgeographic.com/science/article/alice-ball-leprosy-hansens-disease-hawaii-womens-history-science (accessed Jan 8, 2021).

Chapter 6

The Importance of Mentorship and Science Outreach to the Next Generation

Darryl A. Boyd*

Science Made Simple LLC, Fort Washington, Maryland 20744, United States
***Email: drboydthechemist@gmail.com.**

The journey that led Dr. Darryl Boyd into a career in science, and ultimately into STEM (Science, Technology, Engineering and Mathematics) entrepreneurship, is detailed. His recollections of people, programs and events that led to his success in the scientific realm clearly show that exposure to science at an early age, and mentorship from science educators can have a positive and profound impact on a child's desire to pursue science as a career. Though Dr. Boyd has had both men and women scientific mentors, this chapter emphasizes the positive, beneficial mentorship he has received from women in science.

Parental Guidance

I was born in the city of Kalamazoo, in the southwest corridor of the state of Michigan. It was in Kalamazoo, roughly a decade earlier, that my parents, Anthony and Ernestine Boyd, met while they were in college at Western Michigan University (WMU). They were deeply in love and truly lived out the vows they made to one another on their wedding day. They were also devout practicing Christians of the Pentecostal denomination who raised me and my siblings in that faith. I was the third of four children born to this union (Figure 1). My siblings are Marlon (the oldest), Courtney (the only girl) and Nicholas (the youngest). Our parents married just prior to my dad's college graduation when he earned a Bachelor of Science in Sociology. My mother was a pre-med student but did not finish college after they wed. As was common in the 1970's my father desired to work so that my mother did not have to work outside the home. My mother was ambitious and knew she could contribute to the household in more ways than being the stereotypical stay-at-home wife. Shortly after the birth of my brother Marlon, she took on a job as a bank teller at First of America Bank (now PNC) which was headquartered in Kalamazoo. She would ultimately go on to work for more than thirty years at the bank, holding various management positions and becoming an invaluable asset to the company. It turned out that her job at the bank was more stable than my father's jobs, despite his college degree. Thus, my mother's drive and desire to contribute in the household money-making sustained our family countless times when my father was laid off. Tragically, my father died of pancreatic cancer when I was only nine years old. My mother then became the only source of income for herself and her four grade-school-aged children. At the time of my father's death, my mother was a supervisor at the bank, poised to move up in management. Without her career ambition earlier on in their marriage, we would have been in a very precarious financial situation upon my father's unexpected and untimely death. Through this circumstance, from a very early age I knew the strength of women, especially Black women, and how driven, capable and successful they could be.

Figure 1. The Boyd Family in 1985. Front row (l-r) Darryl Boyd, Ernestine Boyd, Nicholas Boyd; Back row (l-r) Anthony Boyd, Marlon Boyd, Courtney Boyd. (Photo courtesy of Dr. Darryl Boyd.)

My father was an excellent dad and husband. Following college, he worked as a counselor at the local jail, but most of my life, he worked at National Water Lift (now Parker-Hannifin) putting together and painting portions of airplanes. He was like a walking book of knowledge. He seemed to know a little something about almost everything. His death dealt a terrible blow to our family. But the things my parents built together in 15 years of parenting, my mother continued after his passing. From attending the Pow-Wow's every year at WMU to learn about Native American culture, to teaching us about religion and differences between our faith and others, my parents sought to expand our worldview, cultivate compassion and generate curiosity. My mother was, and still is, a living example of perseverance and strength.

Early Hands-On Science Experiences

Beginning at a very young age, our parents also introduced us to science. In the early 1970's, there was a man by the name of Dr. LeRoy R. Ray Jr., who launched the "Hands-On Science Training Program" at WMU. His advanced degrees were in science education, but his undergraduate degrees were in biology and biochemistry (*1*). His program was designed to introduce science to grade-school children, particularly to Black children. On Saturdays during the traditional public school calendar year, WMU hosted scores of students as a part of the Hands-On Science Training Program (*2*). Our parents would drop us off at Wood Hall on WMU's campus, and there we would have a general opening session before our classes began for the day. At the beginning of each session Dr. Ray would stand in front of the scores of students and say in a loud, stern voice, "Boys and girls!" Then we would respond, "Yes!" Then he'd say, "Girls and boys!" And we'd respond, "Yes!" He'd finish by asking, "Are we scientists?!" And with great jubilance and enthusiasm, all of us children would exclaim, "Yes!" With that consistent back-and-forth exchange, Dr. Ray was letting us know that we could all be scientists, and that science was exciting. Also, by saying "boys and girls," and following it up with "girls and boys," he was aiming to put all of us on level footing no matter if we were a boy or a girl. Science was for everyone.

When the opening session ended, the students would go to their classes. There were four total class options (Biology, Chemistry, Physics, Computer Science), with three classes taken on a given Saturday, and one class rotated out during each successive week. For example, on the 1st Saturday of the month, my class schedule might have been Biology, Physics and Computer Science. The following Saturday, my class schedule might be Physics, Computer Science and Chemistry. On the 3rd Saturday, my schedule would then be Computer Science, Chemistry and Biology. Then the cycle would start over again. The pattern went on like this throughout the school year. The classes would be led by scientists from various institutions in the area, including WMU and the nearby UpJohn Pharmaceutical Company (now Viatris) (*3*), who volunteered their time to us children.

Each year we also had the year-long "mystery seed" competition. At the beginning of the program year, all of the students were given seeds of unknown identity. We were to grow the seeds into plants in order to identify them. At the end of the school year, judges determined who accurately identified their seeds, and who grew their plants the best. This competition gave us the opportunity to learn about plants and was a great segue into the summer session of the Hands-On Science Training Program. During the summers, Dr. Ray continued the program on his farm in Bloomingdale, Michigan. Parents would again drop their children off at Wood Hall at WMU, and then a van would take the students out to Dr. Ray's farm. On the farm there was a huge barn, farm animals, a man-made pond, and a plot of land marked-off for each of the students to have their own individual garden. There was also a building that housed a classroom where all of the teaching took place. We learned about agriculture, botany, the biology of plants, farm animals, and how to grow a garden. Once again, the classes were taught by scientists from various area institutions, such as WMU, Michigan State University and the UpJohn Company. Perhaps the best part about working out at the farm was that at the end of the summer, we could go to our individual garden plots, and Dr. Ray would encourage us to harvest anything that we were able to grow. Although I never did well in the mystery seed competition, I always grew a wonderful garden.

Of course, no program of this magnitude can be successfully run by a single individual. There were a number of women and men, scientists and staff, who volunteered their time to the program each year. Ms. Sherrie Fuller was one consistent presence at the program throughout my time in it who truly stands out. Ms. Fuller was absolutely amazing, and she was Dr. Ray's right-hand person for the Hands-On Science Training Program. She was a lead facilitator for the program at WMU during the school year sessions, and the primary driver of the van that took us to the farm during the summer sessions. She was the bridge between the students and all of the staff that worked with the program. On those 35-minute van rides from WMU out to the farm, she would carry on conversations with us on a wide range of topics in order to give us insight into the world around us. To this day, she continues to foster the personal and academic growth of BIPOC (Black, Ingenious, people of color) students, having recently held the position of Assistant Director of Multicultural Affairs at WMU (*4*).

Multiple events and occurrences in my youth factored in to my pursuit of a career in science. This included the classes that I took as a part of the Hands-On Science Training Program. I do not recall if I had a favorite course amongst the ones offered by the program, but I do recall many of the demonstrations that the instructors presented in class. For example, one time the chemistry instructor asked us a series of questions about three burning candles that he situated in the front of the classroom. At the end of the demonstration he took one of the candles, tossed it in the air, caught it in his mouth and ate it, but did not explain why or how it was ok for him to eat. I was in disbelief. For years, I thought about that demonstration and tried to figure out how he was able to eat that candle. It was not until many years later that I became familiar with the "edible candle" demonstration, in which oily foods like pecan nuts are lit like a candle wick after being placed in the center of a peeled banana. Those types of demonstrations intrigued me, and played an important role in my interests in science.

Many years after I graduated high school, I was curious about what became of the program. While searching online, I found an Associated Press article about the program entitled "Curiosity Nurtured in Black Youngsters" that ran in the *Los Angeles Times* in October of 1993 (*2*). To my surprise, there was a passage in the article that quoted my sister, who had been interviewed by a reporter for the article one day when we were out at Dr. Ray's farm. In the article, my sister was quoted as saying, "Dr. Ray's farm opened my knowledge for science" (*2*). The article also pointed out that my sister was heavily influenced by the program, and that she desired to become a pediatrician one day (*2*). Though my sister ultimately did not pursue a career in science or medicine, it is clear from the quotes she gave the reporter that participating in the Hands-On Science Training Program inspired her.

After many years in the Hands-On Science Training Program, my mother placed me and my younger brother in other science programs, including the Kalamazoo Math and Science Pre-college Engineering Program (KMASPEP) at the Kalamazoo Area Mathematics and Science Center (KAMSC), and the Science In The City program. KMASPEP was similar to the Hands-On Science Training Program in that most of the courses were led by instructors from the schools and colleges in the Kalamazoo area, and there were both classroom as well as in-the-field hands-on aspects to the program. Science In The City was a program that allowed us to do hands-on science and engineering while visiting various colleges in the state of Michigan, including Central Michigan University in Mount Pleasant, MI, Lawrence Technological University in Southfield, MI, and Madonna University in Livonia, MI. These programs also focused their primary attention on encouraging BIPOC students to take an interest in fields that fall under what is now called STEM (Science Technology Engineering & Mathematics).

Dr. Ray passed away in 2013 at the age of 82 (*1*). A few years prior to his passing, while I was working on my Ph.D. at Purdue University, I received a call from him while I was at my desk in the lab. Apparently, he had a chance encounter with my sister while they were both out shopping in Kalamazoo and she gave him my number after telling him of my pursuit of a Ph.D. in Chemistry. After a brief discussion of the work I was doing in the lab, he told me that he was going to take some credit for the scientific career path I'd taken. I let him know that he was well within his rights to do so because the Hands-On Science Training Program gave me a very positive, engaging early introduction into the world of science.

Women Scientific Mentors

Beginning with my father, there have been many men in my life that served as role models, and who helped me as I have traveled this scientific research path. Similarly, beginning with my mother, I have had numerous women help me to become the scientist that I am today. That includes many of my grade school teachers. With the exception of my 3rd grade teacher, Mr. Thomas McCormick, all of my teachers in kindergarten through 6th grade were women. In order, starting with kindergarten, they were Mrs. Lois Hoekstra, Mrs. Carol Berry, Mrs. Auga Davis, Mrs. Tina Echols, Mrs. Ann Nieboer, and Mrs. Sara Glendening. Their names are forever etched in my memory, as are many lessons each of them taught me and my classmates.

I finished my grade school years at Kalamazoo Central "World Class" High School (KCHS). When I entered high school there was one lady in particular who was very helpful in guiding me through the STEM fields. Her name was Dr. Brenda Earhart, and she was a faculty member at the aforementioned KAMSC. Dr. Earhart was already a familiar face to me because she founded and led the KMASPEP program that I was a part of some years earlier prior to high school. KAMSC was a high school math and science institution that required all of the students to score well enough on an exam in order to enter the school. If you passed the exam and enrolled, half of your school day would be spent at your home school institution (i.e. KCHS), and the other half of your school day would take place at KAMSC, where you took math and science courses. The course credits earned at KAMSC counted toward your high school diploma the same as if you took math and science courses at your home school.

Dr. Earhart was a Black woman who strongly encouraged young Black students in the Kalamazoo area to take the KAMSC entrance exam, and to enroll if they scored high enough on the exam. Upon taking the exam, I was accepted as a student at KAMSC along with a few other friends. The curriculum there was rigorous and quite difficult. Many of the courses were taught at a college level, and there were some students at KAMSC who had even tested out of some of those difficult KAMSC courses and were taking math and science courses at local colleges. Because of the rigor, everyone there needed a support system and guidance. For the Black students at KAMSC - of which there were only a few - Dr. Earhart was a major part of that support system. Dr. Earhart, who had a Master of Science in Chemistry and a Doctorate in Educational Leadership, would stay many hours after school with us helping us to succeed in all of our classes at KAMSC. She was also a strong advocate for us whenever we faced challenges with staff or other students. Dr. Earhart ultimately became the Director of KAMSC, before retiring in 2011 (5).

While at KAMSC during high school, I was able to take my first formal chemistry courses. The courses were General Chemistry, taught by Mrs. Carolyn Virkhaus, and Organic Chemistry taught by Ms. Cheryl Hach, both women scientists. I specifically recall an instance in which Mrs. Virkhaus was asking us to tell her the name of compounds based on the chemical formulas she wrote on the chalk board. She quizzed the class on a few compounds and I was able to correctly name most of them before the rest of my peers. Then she wrote "$KC_2H_3O_2$" on the board and without hesitation I blurted out "potassium acetate." She looked at me and said, "You really have a knack for nomenclature." The truth is, nomenclature was second nature to me at that point because it was covered in those first science programs my mother and father placed me in prior to high school.

Having both an interest in science and a knack for the subject matter, upon graduation from high school, I chose to pursue chemistry as my major when I enrolled in college at the University of Michigan. During my final years at Michigan, I was able to perform undergraduate research in Professor Vincent Pecoraro's inorganic chemistry lab, and also in the environmental engineering lab led by Professor Linda Abriola. I learned many valuable lessons from both professors while working in their respective labs. Both led with an optimism and a positivity that is sometimes lacking in science mentoring and leadership roles. I was fortunate to take two inorganic chemistry courses with Professor Pecoraro in addition to working in his lab. I was honored to be working in Professor Abriola's lab when she was elected to the National Academy of Engineering (NAE) for her groundbreaking research in liquid contaminants (*6*). Her students, who highly revered and respected her and her research, emphasized to me that Professor Abriola's election to the NAE was long overdue. Shortly after my time in her lab, she left the University of Michigan to become the Dean of the School of Engineering at Tufts University outside of Boston, MA.

Upon receiving my Bachelor of Science in Chemistry degree from the University of Michigan, I enrolled in the Chemistry Department at Purdue University in pursuit of a Ph.D. in biochemistry. Losing my father to cancer was a major factor in my decision to pursue the Ph.D. at Purdue. At the time, there were a number of faculty members in the Chemistry Department doing research on cancer. I joined the lab of Professor Christine Hrycyna, whose work had implications in pancreatic cancer. Prior to my arrival her lab comprised all women, and though two other men joined the lab along with me, there was an extended period of time that I was the only man in the lab upon their departures. It was in Professor Hrycyna's lab that I learned the value of giving excellent research presentations. Professor Hrycyna was very strict about how presentations should be given in order to convey the desired message, and she trained all of her students accordingly. I credit not only Professor Hrycyna but also the women that worked in her lab with improving the quality of my research presentations. Just hours prior to my first presentation in front of Professor Hrycyna and the group, I showed my slides to some of my lab-mates. Upon reviewing the slides, they were adamant that I needed to improve the visual content on each slide. I told them that I was not as worried about the look of the slides because I could explain all of the content if anyone had questions. Each of the women then immediately responded with great clarity that the visual presentation is as important as the understanding of what is being presented. I took what they said to heart, and from that point forward, my oral research presentations have improved tremendously.

Although I learned how to perform polymerase chain reaction (PCR), site-directed mutagenesis and protein purification in Professor Hrycyna's lab (along with sharp visual presentation skills), I lacked a stronger foundation in the fundamentals of biology, and completing my Ph.D. in a biochemistry lab was not the best route for me. So I left Professor Hrycyna's lab with a thesis and Master of Science degree. I then joined Professor Tong Ren's inorganic chemistry lab where I was far more comfortable with the underlying fundamentals and subject matter. During my time in Professor Ren's lab, I performed a tremendous amount of organic and inorganic syntheses of molecules with a focus on electron transfer. I also learned how to obtain and interpret electrochemical data from the molecules I synthesized using techniques such as cyclic voltammetry. With Professor Ren's guidance and support, I was able to complete my Ph.D. in inorganic chemistry in his lab in three years, leaving with three first-author publications (*7–9*).

Following my Ph.D. work, I landed a postdoctoral position as a National Research Council Research Associate. The work was conducted at the United States Naval Research Laboratory (NRL), and my postdoctoral advisor was Dr. Frances S. Ligler. My knowledge of inorganic compounds and organometallic syntheses was the central skill for which Dr. Ligler hired me. Like Professor Abriola, she was a member of the National Academy of Engineering, and she was a pioneer in the development and research of biosensors. Her style of leadership was to give her trainees a framework to work within, but to allow those trainees to build their own research profile within that framework. She also demonstrated that it is perfectly fine to lack knowledge, even within your field of training, as long as you are willing to put in the work to fill that knowledge gap. Under her tutelage my passion for independent research grew immensely. My skills as a researcher, including scientific writing and building a research network, were also sharpened during my time as her postdoctoral associate. The work I did in Dr. Ligler's lab resulted in five first-author research publications and two additional co-authored reviews, all in high-impact peer-reviewed journals (*10–12*). Upon completion of my postdoctoral research with Dr. Ligler, I was hired on as a full-time employee at NRL as a Research Chemist in the Optical Sciences Division.

Independent Polymer Research

My career research journey includes time in inorganic chemistry, biochemistry, and polymer science laboratories. Of those areas of research, the work that I am most commonly associated with is within the polymer science realm. I began working with polymers during my postdoctoral tenure under the direction of Dr. Ligler at NRL. The majority of that work focused on utilizing thiol-ene and thiol-yne coupling chemistries to make polymer fibers that had non-round cross-section shapes (*10, 11*). Thiol-coupling methodologies depend on the ability of sulfur-containing thiol groups to couple with available electrons from unsaturated carbon-carbon bonds (*13, 14*). This coupling can be initiated by ultraviolet (UV) irradiation, making these reactions useful in an array of processes and technologies (*13, 14*). We used custom-designed microfluidic channels, coupled with UV irradiation, to make the fibers. Using microfluidic approaches to make the fibers instead of other common methods, such as electrospinning, afforded us more control over the cross-section shape of the fibers (*15*). The scope of this thiol-coupling work was later expanded to include the development of polymer nanocomposites (*12, 16, 17*).

In addition to serving as my introduction to polymer science, thiol-coupling research was my first foray into sulfur-based chemistry. Sulfur is an abundant element with numerous properties that make it useful in a wide-range of research fields including biochemistry, inorganic chemistry and polymer science (*18*). Not only is sulfur abundant, it is also ubiquitous, being found in such disparate locations as amino acids in the human body, as well as in the atmosphere of Earth's planetary neighbor Venus (*19*).

Concurrent with the thiol-coupling work I do, I have worked with sulfur in its elemental form in order to develop polymers that are optically useful. These efforts utilize a newly developed polymerization method called inverse vulcanization (*20*). In traditional vulcanization, sulfur is the crosslinker between backbone molecular chains that are primarily made of carbon (*18*). Inverse vulcanization reverses these roles, meaning that primarily carbon-containing molecules are used to crosslink backbone chains made up of sulfur (*20*). Unique properties of these types of polymers include unusually high refractive indices ($n > 1.8$), and optical transmission that extends from the visible into the infrared. In practice, these types of optical materials could be used as lighter and cheaper replacements for materials that are currently used in visual detection devices like night-vision goggles.

Along with my colleagues in the Optical Sciences Division at NRL, we joined this field of research to develop infrared-transmissive optical polymers. Our team at NRL is credited as the first to include more than one element in the backbone of inverse vulcanized polymers. Specifically, we incorporated both sulfur and selenium in the backbone of polymers that had 1,3-diispropenyl benzene (DIB) as the crosslinker (*21*). We are also credited with being the first group to use an organometallic molecule as the crosslinker in an inverse vulcanization reaction (*22*). To the best of our knowledge, we were also the first group to demonstrate useful long-wave infrared (LWIR) transmission through bulk polymers (thickness > 1 mm) (*22*). This LWIR transmission discovery has implications in polymers and optical materials that go beyond the inverse vulcanization polymer sub-field (*23*). Overall, our inverse vulcanization research has resulted in multiple publications and patent disclosures, and I have been honored with a number of awards and honors due to these efforts, including being named to the 2018 class of *Chemical & Engineering News*' "Talented 12" (*24*).

Science Entrepreneurship

Since joining the student chapter of the National Organization of Black Chemists & Chemical Engineers (NOBCChE) as an undergraduate student at the University of Michigan in the year 2000, I have participated in many community STEM outreach activities. Initially, my efforts in this arena occurred as a part of my involvement with organizations like NOBCChE. When I was pursuing my Ph.D., I began to focus some of my outreach efforts on my younger relatives, who thought that it was so "cool" that I was a scientist. Among these relatives was one of my nieces who, beginning around age 4, would often remark that she wanted to be a scientist like "Uncle Darryl." Several high school and early college-aged relatives also expressed great interest in the scientific career path I chose. I noted in discussions with some of them that they realized they could be scientists because they knew that I, a Black man who was their relative, was a scientist.

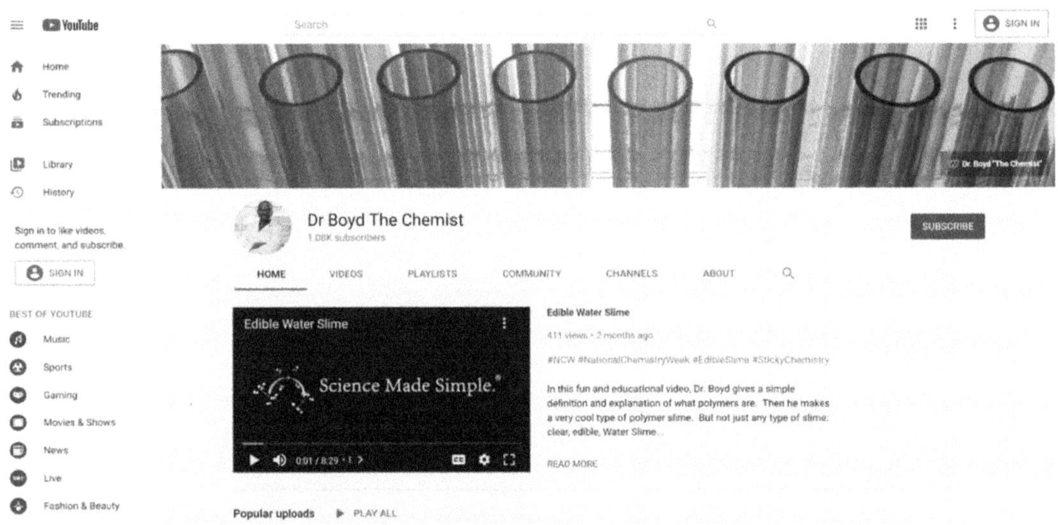

Figure 2. Screenshot of the "Dr. Boyd The Chemist" YouTube Channel home page.

Knowing that there were few Black scientists with a public platform for youths like my young relatives to look up to, I decided to expand my science outreach efforts. This included volunteering to teach science lessons at elementary schools, participating as a judge at grade-school and college level science fairs, tutoring chemistry for high-schoolers and college undergraduates, and continuing my involvement in science outreach activities that were facilitated by my membership in various professional organizations. I launched a website and companion YouTube channel in which I create and curate educational science content (Figure 2). After years of performing these more extensive outreach efforts, I decided to start a company through which to perform and disseminate materials. The company, "Science Made Simple LLC," has emphasis in three areas: 1) scientific educational outreach, 2) public scientific communication and 3) groundbreaking scientific research. The outreach sector of the company encompasses the hands-on and interactive STEM outreach efforts that I have already engaged in for many years. The public communication aspect focuses on opportunities to speak to the public about science in a formal, yet engaging way. The research angle affords the opportunity to pursue research ideas that fall outside of the scope of my job at NRL.

Science Made Simple was co-founded with my wife Sara. My wife is Puerto Rican and was raised on the island. We met while we were at Purdue University, where she earned a degree from Purdue's famed Krannert School of Management. Since her graduation, she has established herself as a much sought-after sales person and as an entrepreneur in career consulting and coaching. Although my wife is not a scientist, she often travels with me, performing and accurately explaining the science demonstrations that we perform for diverse audiences. The volume of science understanding she has gained in the years since we met impresses me. I think her engagement with science as a person who works in a totally different field serves as a demonstration of just how accessible and interesting science can be to anyone.

Chapter Summary

Effective mentoring was absolutely essential to my success as a chemist. Furthermore, I was mentored by several amazing women, including my mother. I know the strength of women - especially Black and Latina women - and how driven, capable and successful they are in the field of chemistry. I know the importance and significance of women scientific mentors.

Figure 3. Dr. Darryl Boyd (left) with his wife and daughter in October 2020. (Photo courtesy of Dr. Darryl Boyd.)

In many ways, Science Made Simple targets young people, especially those who are historically underrepresented in the STEM fields, which aligns with the key goals of Dr. Ray's Hands-On Science Training Program. The goal is to give the children a fun and solid science foundation, while encouraging them that they too can be scientists. Through Science Made Simple, I hope to inspire children to take an interest in science even if it is not the focus of their ultimate career path. This is a personally important mission for me because of the interest of so many of my younger relatives in science.

My wife and I recently became parents to a beautiful daughter (Figure 3). Our daughter is both Black and Latina, and unfortunately, there aren't many Black or Latina women in STEM (*25*, *26*). Due to the small percentage of women and people of color in STEM, I want my daughter to see that she too can pursue a career in science if she so desires. If she chooses another career path, I hope that seeing her mother actively engaging with science exemplifies that a person does not have to be a career scientist in order to be well-versed in basic science concepts, or to appreciate the value of science in society. I want her and the young people I engage through Science Made Simple to know that science is for everyone!

My polymer research discussed within this chapter provides yet another opportunity for engagement with advanced chemistry majors in the classroom. Thus, the appendix of this chapter includes a lesson plan based on a recent publication in the high-impact journal *ACS Macro Letters*. In addition, general chemistry concepts are also included for first-year students aligned with concepts focused on nomenclature.

Appendix

Sample Teaching Chemistry Assignment: The Chemistry of Sulfur

Literature Reading Assignment I

Read the following article before class:

> Torrice, M. "Darryl Boyd: Polymers pioneer is stitching sulfur into molecular strands to create advanced optics," *Chemical and Engineering News*, https://cen.acs.org/materials/polymers/Darryl-Boyd/96/i33.

Answer the following questions for class discussion on Dr. Darryl A. Boyd:

1. Where did he earn his Ph.D. in chemistry?
2. What are Dr. Boyd's favorite television shows?
3. What is his favorite fictional element?
4. Why is sulfur added to eyeglass lenses?

General Chemistry Concepts: Nomenclature

1. What is the definition of a polyatomic ion?
2. In this chapter, it is stated that the formula $KC_2H_3O_2$ is called potassium acetate. What is the overall charge on the acetate ion?
3. Given the name, write the chemical formula for the following compounds:

    a. Calcium acetate
    b. Magnesium phosphate
    c. Sodium sulfate

4. Given the chemical formula, write the name for the following compounds:

    a. $Mg(C_2H_3O_2)_2$
    b. NH_4Cl
    c. NH_4NO_3

5. Dr. Boyd mentioned that sulfur is found in amino acids in the human body. Which amino acid(s) in the human body contain(s) sulfur?

Literature Reading Assignment II

Read the following article before class:

> Boyd, D.A.; Nguyen, V.Q.; McClain, C.C.; Kung, F.H.; Baker, C.C.; Myers, J.D.; Hunt, M.P.; Kim, W.; Sanghera, J.S., "Optical Properties of a Sulfur-Rich Organically Modified Chalcogenide Polymer Synthesized via Inverse Vulcanization and Containing an Organometallic Comonomer," ACS Macro Letters, 2019, 8, 113–116.

Answer the following questions for a class discussion:

1. Briefly describe the significance of this study. Why is this research important?
2. What is inverse vulcanization?
3. (a) What are refractive indices? (b) Why did the authors use refractive index data in the study? (c) Explain the refractive index data in Table 1 of the paper.
4. What other laboratory techniques were utilized to characterize the sulfur-based polymers described in the paper?
5. Summarize the conclusions of the paper.

Creative Poster for a Gallery Walk

You will work in teams to prepare a creative poster based on the peer-reviewed article co-authored by Dr. Boyd and his colleagues. You will use large 3M Post-It paper (25 x 30) and colorful markers to complete the team poster. The poster must include responses to the five discussion questions and sketches of the chemical structures of the sulfur-based polymers described in the paper. You will use a rubric prepared by the class to evaluate the team posters for the gallery walk.

References

1. Barron, J. *Dr. Leroy Ray Jr. obituary*. https://wmich.edu/news/2013/02/4919 (accessed Nov 28, 2020).
2. Perlman, L. Curiosity Nurtured in Black Youngsters. *LA Times*, https://www.latimes.com/archives/la-xpm-1993-10-03-mn-41639-story.html (accessed Nov 28, 2020).
3. Sabatini, P. Mylan completes merger with UpJohn to form Viatris. *Pittsburgh Post-Gazette*, https://www.post-gazette.com/business/healthcare-business/2020/11/16/Mylan-merger-Upjohn-Viatris-bresch-coury-Pfizer-cost-cuts-generic-drugs/stories/202011160111 (accessed Dec 16, 2020).
4. Barron, J. *Longtime employees to be inducted into 25 Year Club*. https://wmich.edu/news/2014/11/20077 (accessed Nov 28, 2020).

5. *Kalamazoo Area Mathematics & Science Center, Former Staff Profile for Dr. Brenda Earhart.* https://kamsconline.com/staff/former/DrBrendaEarhart.aspx (accessed Nov 26, 2020).
6. *National Academy of Engineering, Linda M. Abriola Biography.* https://www.nae.edu/155235/Linda-M-Abriola (accessed Nov 26, 2020).
7. Boyd, D. A.; Fanwick, P. E.; Ren, T. New Diruthenium Complexes Formed Via Modification With 1,1′-Ferrocene Dicarboxylic Acid. *Inorg. Chim. Acta* **2011**, *370*, 198–202.
8. Boyd, D. A.; Cao, Z.; Song, Y.; Wang, T.; Fanwick, P. E.; Crutchley, R. J.; Ren, T. Diruthenium Compounds Bearing Equatorial Fc-containing Ligands: Synthesis and Electronic Structure. *Inorg. Chem.* **2010**, *49* (24), 11525–11531.
9. Boyd, D. A.; Crutchley, R. J.; Fanwick, P. E.; Ren, T. Fc-Fc Electronic Interaction through Equatorial Pathways of a Diruthenium Core. *Inorg. Chem.* **2010**, *49* (4), 1322–1324.
10. Boyd, D. A.; Shields, A. R.; Naciri, J.; Ligler, F. S. Hydrodynamic Shaping, Polymerization, and Subsequent Modification of Thiol Click Fibers. *ACS Appl. Mater. Inter.* **2013**, 5, 114–119.
11. Boyd, D. A.; Shields, A. R.; Howell, P. B.; Ligler, F. S. Design and Fabrication of Uniquely Shaped Thiol–ene Microfibers Using a Two-Stage Hydrodynamic Focusing Design. *Lab Chip* **2013**, *13*, 3105–3110.
12. Boyd, D. A.; Naciri, J.; Fontana, J.; Pacardo, D. B.; Shields, A. R.; Verbarg, J.; Spillmann, C. M.; Ligler, F. S. Facile Fabrication of Color Tunable Film and Fiber Nanocomposites via Thiol Click Chemistry. *Macromolecules* **2014**, *47*, 695–704.
13. Hoyle, C. E.; Bowman, C. N. Thiol-Ene Click Chemistry. *Angew. Chem. Int. Ed.* **2010**, *49*, 1540–1573.
14. Lowe, A. B. Thiol-yne 'click'/coupling chemistry and recent applications in polymer and materials synthesis and modification. *Polymer* **2014**, 55, 5517–5549.
15. Daniele, M. A.; Boyd, D. A.; Ligler, F. S.; Adams, A. A. Microfluidic Strategies for Design and Assembly of Microfibers and Nanofibers with Tissue Engineering and Regenerative Medicine Applications. *Adv. Healthcare Mater.* **2015**, 4, 11–28.
16. Boyd, D. A.; Bezares, F. J.; Pacardo, D. B.; Ukaegbu, M.; Hosten, C.; Ligler, F. S. Small-Molecule Detection in Thiol-Yne Nanocomposites vis Surface-Enhanced Raman Spectroscopy. *Analytical Chemistry* **2014**, *86*, 12315–12320.
17. Stewart, M. H.; Susumu, K.; Oh, E.; Brown, C. G.; McClain, C. C.; Gorzkowski, E. P.; Boyd, D. A. Fabrication of Photoluminescent Quantum Dot Thiol-yne Nanocomposites via Thermal Curing or Photopolymerization. *ACS Omega* **2018**, 3, 3314–3320.
18. Boyd, D. A. Sulfur and its Role in Modern Materials Science. *Angew. Chemie* **2016**, *55* (50), 15486–15502.
19. Esposito, L. W.; Winick, J. R.; Stewart, A. I. Sulfur dioxide in the Venus atmosphere: Distribution and implications. *Geophysical Research Lett.* **1979**, *6* (7), 601–604.
20. Chung, W. J.; Griebel, J. J.; Kim, E. T.; Yoon, H.; Simmonds, A. G.; Ji, H. J.; Dirlam, P. T.; Glass, R. S.; Wie, J. J.; Nguyen, N. A.; Guralnick, B. W.; Park, J.; Somogyi, Á.; Theato, P.; Mackay, M. E.; Sung, Y.-E.; Char, K.; Pyun, J. The Use of Elemental Sulfur As An Alternative Feedstock For Polymeric Materials. *Nat. Chem.* **2013**, 5 (6), 518–524.
21. Boyd, D. A.; Baker, C. C.; Myers, J. D.; Nguyen, V. Q.; Drake, G. A.; McClain, C. C.; Kung, F. H.; Bowman, S. R.; Kim, W.; Sanghera, J. S. ORMOCHALCs: Organically Modified Chalcogenide Polymers For Infrared Optics. *Chemical Communications* **2017**, 53, 259–262.

22. Boyd, D. A.; Nguyen, V. Q.; McClain, C. C.; Kung, F. H.; Baker, C. C.; Myers, J. D.; Hunt, M. P.; Kim, W.; Sanghera, J. S. Optical Properties of a Sulfur-Rich Organically Modified Chalcogenide Polymer Synthesized via Inverse Vulcanization and Containing an Organometallic Comonomer. *ACS Macro Letters* **2019**, *8*, 113–116.
23. Kleine, T. S.; Lee, T.; Carothers, K. J.; Hamilton, M. O.; Anderson, L. E.; Diaz, L. R.; Lyons, N. P.; Coasey, K. R.; Parker, W. O., Jr; Borghi, L.; Mackay, M. E.; Char, K.; Glass, R. S.; Lichtenberger, D. L.; Norwood, R. A.; Pyun, J. Infrared Fingerprint Engineering: A Molecular-Design Approach to Long-Wave Infrared Transparency with Polymeric Materials. *Angew. Chem. Int. Ed.* **2019**, *58* (49), 17656–17660.
24. Torrice, M. Darryl Boyd: Polymers Pioneer is Stitching Sulfur into Molecular Strands to Create Advanced Optics. *Chemical and Engineering News*, https://cen.acs.org/materials/polymers/Darryl-Boyd/96/i33 (accessed Nov 11, 2020).
25. Salman, J. Just 3% of scientists and engineers are Black or Latina women. Here's what teachers are doing about it. *The Hechinger Report*, https://hechingerreport.org/just-3-of-scientists-and-engineers-are-black-or-latina-women-heres-what-teachers-are-doing-about-it/ (accessed Dec 15, 2020).
26. *National Science Foundation – Women, Minorities, and Persons with Disabilities in Science and Engineering: 2017 Digest*, https://www.nsf.gov/statistics/2017/nsf17310/static/downloads/nsf17310-digest.pdf (accessed March 23, 2021).

Chapter 7

Organic Chemistry, Consulting, and the STEM Butterfly Effect

Pamela Leggett-Robinson*

PLR Consulting, Hampton, Georgia 30228, United States

***Email: pamela@plrconsultingservices.org.**

This chapter will focus on my chemistry journey navigating a career in academia, teaching, and establishing a consulting firm to address equity and inclusion in STEM (Science, Technology, Engineering and Mathematics) education initiatives. Furthermore, I will address barriers to success in the chemical sciences and the importance of effective mentoring to inspire the next generation of chemists.

Early Years: Undergraduate Chemistry Education

After graduating from high school, I attended an R1 institution in the state of Florida. I was a pharmacy major and needed to enroll in biology and chemistry classes for my degree. Reflecting on my undergraduate experiences at this institution really emphasizes the importance of science identity and sense of belonging for BIPOC (Black, Indigenous, and people of color) communities. These experiences would later serve as my motivation to address equity and inclusion in the STEM fields. STEM belonging or sense of belonging is defined as an individual's feelings of acceptance and integration, more specifically, whether he or she feels included in a particular environment (*1*).

In order to enroll in general chemistry courses, all students were required to pass an entrance exam. Mathematics was a challenging subject for me as an undergraduate and it was a barrier in passing the exam. I did eventually pass the examination. Unfortunately, after enrolling in this class, I never performed well on the examinations. Moreover, I experienced more challenges and significant isolation in my biology laboratory course. Eventually, I made the difficult decision to leave the institution, return home and enroll at a local community college. Eventually, I enrolled at Georgia State University to complete my undergraduate degree. I have shared my difficult experiences as an undergraduate student previously (*2*), so I will not provide significant detail in this volume. However, those challenging life experiences were critical to my current STEM advocacy work.

As an undergraduate, at Georgia State University, I had the opportunity to engage in research as a Ronald E. McNair Scholar and gained hands-on chemistry laboratory experiences. I am truly thankful to Dr. C.T. Cummings, Dr. Davon Kennedy, and Dr. Alfons Baumstark for their guidance and support to allow me the opportunity to participate in the McNair Program. My research project focused on synthetic organic chemistry, where my advisor expected laboratory group members to synthesize and characterize new compounds. Honestly, I don't recall specific experimental details of my research project, but I do remember the mentoring that I received from Dr. Lucjan Strekowski.

Although I was looking forward to graduating and attending graduate school, I was also planning my wedding. Dr. Strekowski knew this and called me into his office one afternoon. He asked me which graduate schools I had considered for further study. I shared with him that I planned to apply to the University of Michigan and the University of North Carolina at Chapel Hill for graduate study. Dr. Strekowski quickly reminded me of my lack of discipline in critically reading the chemistry literature and my pending nuptials. "These two things make for a bad graduate student," he stated. He continued by sharing that typically when women get married and have children this can delay earning a doctoral degree. He encouraged me to consider a smaller institution and obtain a master's degree first. "That way, you will leave the institution with something," he stated. I was so angry. Dr. Strekowski assumed I wanted children, and assumed I was unable to be a wife, mother, and graduate student simultaneously. Nevertheless, I needed his recommendation. Upon reflection of this conversation, Dr. Strekowski provided me with great and honest advice about pursuing graduate work. Thus, I enrolled at Tennessee Technological University in Cookeville, TN to pursue my graduate degree in chemistry.

Graduate Study at Tennessee Technological University

Graduate school at Tennessee Technical University (TTU) was an interesting experience. I was a teaching assistant (TA) and enjoyed research focused on inorganic chemistry. My advisor was a new assistant professor, Dr. Edward Lisic, who had more faith in my laboratory skills than I did. His level of excitement about research motivated me to take ownership of my project. I enrolled in several courses including organic spectroscopy and nuclear medicine. Moreover, I saw tremendous growth in my skills and abilities as a bench chemist and instructor. I am truly thankful to Dr. Tom Furtsch, who mentored me to become an effective TA and develop a stronger presence in the classroom. My goal was to make chemistry fun and connect with my students.

TTU is a PWI (Predominately White Institution) and the Black student population was approximately five percent. Although there were only a small number of Black students on campus, I didn't have the experiences of isolation and felt a sense of belonging. I was engaged in social activities and interacted with the other Black students on campus. Moreover, the Black Cultural Center was our safe space, our community.

Through my sorority, I was introduced to Dr. Rubye Torrey, professor emeritus and assistant vice president for research. Dr. Torrey earned her PhD in chemistry (3) from Syracuse University in 1969. Dr. Torrey was the first Black woman I met who earned a Ph.D. in chemistry. She constantly encouraged me to pursue a PhD after obtaining my MS degree. She shared with me her difficult journey in chemistry and was truly an inspiration and mentor for me.

My last semester at TTU, my husband and I found out we were expecting a child. I was so happy that I took the advice from Dr. Strekowski about graduate work. I successfully defended my MS thesis and I was tremendously proud that Dr. Torrey attended my thesis defense. My parents were ecstatic. I was the first in the family to attend college, and I was the first to obtain both a BS and MS degree. I was also thinking about earning my doctoral degree in chemistry.

Between graduating from TTU with my MS degree and enrolling in a doctoral program at GSU, I took a position as an Instructor at North Greenville College (now known as North Greenville University), in Tigerville, SC. I taught math, physical science, and chemistry courses. North Greenville is a private Baptist institution and there was a lack of diversity among the faculty, students and staff during that time. I remember my first day and the look on the Black students' faces. They were so happy to see someone that looked like them on the faculty. There were few Black faculty and staff on campus, and my office became a "safe space" for Black students. One student, Ken Etheredge, was particularly faithful in stopping by my office at North Greenville. He and his friend would often talk about their studies and difficulties of "fitting in" at North Greenville. Our mentee-mentor relationship continued to grow and he stepped in to take over my laboratory course, while I was on maternity leave with my son. Although he was not a STEM major, the skills he learned in my class and in managing the labs, allowed him the opportunity to develop a growth mindset and develop his leadership skills. Ken eventually transferred from North Greenville and completed his degree at a four-year institution. I also left North Greenville to pursue my doctoral degree in chemistry. However, the experience and the relationship had an impact on us both.

Ken, who is now director of finance for the Republic National Distributing Company, recently shared a reflection about his experiences as my mentee at North Greenville. "Many of the lessons I learned from Pam (as a student and an assistant) have benefited me throughout my career and I often reflect on how graceful she handled being the lone minority instructor on campus gaining respect from students and administration," he wrote. Thus, I clearly understood that my role as a mentor is critical for student success.

Doctoral Study at Georgia State University

I enjoyed teaching at North Greenville, but soon realized I needed a Ph.D. to become a tenure-track faculty member. Thus, I reached out to Dr. Al Baumstark at Georgia State, who was still the department chair at GSU and we discussed the rigors of the doctoral program. Upon taking the GRE (Graduate Record Examination) and completing the application process, I was accepted into the program. Dr. Baumstark and I spoke again and determined that his lab would be the best fit for me. During this time, I was separated with two small children, a daughter and son, ages 5 and 3.

As a graduate student, I was required to conduct research, take classes, pass both written and oral qualifying exams, and deliver a research seminar. I thoroughly enjoyed both the research and most of my classes. However, my written qualifying exam was difficult and I did not pass. I begin to question my skills and abilities as a chemist. I remember thinking I was not smart enough. How am I going to get through this program? To my surprise, the department did not dismiss me from the program. Instead, the department provided me with the opportunity to take a more targeted approach in my graduate studies focused on organic chemistry. First, I would work closely with Dr. Keith Pascoe, a brilliant organic chemistry instructor who served as an important mentor to improve my classroom presence and deepen my understanding of organic chemistry. I would sit in Dr. Pascoe's lecture class and grade exams. During the extensive grading sessions, Dr. Pascoe would ask me why the answer was correct for the problems, which enhanced my critical thinking skills. This cognitive apprenticeship approach for my doctoral studies in organic chemistry was absolutely invaluable to me.

Preparing for my research seminar was very difficult. The instructor of the seminar course, Dr. Dabney Dixon was tough! After each preparation session, I felt more and more unsure of myself. Although I had finally passed the written exam, she made me feel as though I still didn't belong in the program. The night before the presentation was perhaps the worst practice session. I left her office seriously contemplating leaving the program after my seminar. Once I arrived at home, I critically reviewed every slide for my presentation until I became very comfortable talking about my research. The next day, I delivered my research seminar and received a great response from the audience and Dr. Dixon. For the first time, she gave me praise for my efforts. I knew then, I belonged in the doctoral program and teaching would be my calling.

As I continued to successfully progress in the doctoral program, I gained more teaching responsibilities in the department. I transitioned from being a TA, and began delivering pre-laboratory lectures for an evening organic laboratory course. In addition, I was awarded with a teaching fellowship and the SREB (Southern Regional Education Board) Dissertation Fellowship. I also volunteered to speak with the McNair Scholars and led STEM-based activities with my daughter's Girl Scout Troop.

My doctoral research under the direction of Dr. Baumstark focused on the organic molecule dimethyldioxirane (Figure 1), which is a powerful stereoselective oxidizing agent. This doctoral research is important to society because dimethyldioxirane is an important reagent used for many synthetic organic chemistry applications such as the oxidation of alkenes to epoxides (*4*).

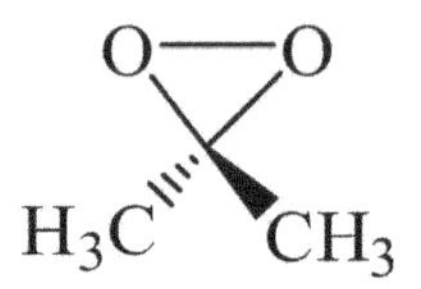

Figure 1. The chemical structure of dimethyldioxirane.

As I continued to complete my doctoral research, Dr. Dixon became an important mentor for me in the department. During the fall semester of my final year, I enrolled in another course taught by Dr. Dixon. This particular course was focused on critically reviewing the literature and developing effective writing skills. Dr. Dixon understood my goals of entering the professoriate and she tailored my classwork to focus on writing grant proposals. This grant writing experience was significant for me later as an assistant professor.

Furthermore, Dr. Dixon offered advice focused on work-life balance such as meal preparation, seeking childcare support when attending scientific conferences, and forming professional networks. Although we had discussed my career pathway in academia, she made me think about the importance of selecting the "right" institution and department. During these crucial conversations about my career path, I discovered my desire to educate STEM students that looked like me. Once my career decision was made, Dr. Dixon picked up the phone and called an alumnus of GSU, Dr. Gregory Pritchett, who was now the Chair of the Department of Chemistry at Tuskegee University, an HBCU (Historically Black Colleges and Universities), in Alabama. I successfully defended my doctoral dissertation (*5*) in 2003 and then joined the faculty at Tuskegee University. This was the beginning of my quest to broaden participation in STEM within the academy.

The Professoriate: The HBCU Experience

HBCUs play a significant role in higher education (*6–8*) and often provide more inclusive and supportive environments for students from BIPOC communities compared to PWIs. In addition, HBCU faculty members provide a strong sense of belonging through institutional cultures of caring and empathy. This type of supportive academic environment facilitates the development of self-efficacy, self-esteem, and self-identity, which validates a student's purpose at the academic institution and in life (*9*).

Moving from graduate school to the academy was an interesting transition. I was excited to be teaching at an HBCU, but nervous nonetheless. The chemistry and biology department chairs, Dr. Greg Pritchett and Dr. Roberta Troy welcomed me and nurtured my academic and professional development as an assistant professor. My first semester, I was introduced to Dr. Derrick Dean, a professor of material science. This would lead to a collaboration with Dr. Dean and my first National

Science Foundation (NSF) funded proposal. Dr. Troy nominated me for a National Institutes of Health (NIH) Steps Toward Academic Research (STAR) Health Disparities Fellowship at the University of North Texas Health Science Center (UNTHSC), for the Faculty Senate, and eventually for the Faculty Senate Chair. My department chair, Dr. Pritchett constantly gave advice for navigating the HBCU landscape. He emphasized the importance of speaking to everyone, being kind and showing empathy towards others. At the end of each semester, Dr. Pritchett would often talk with me about my career trajectory and to seek other academic career opportunities.

Teaching at Tuskegee University was more difficult than I could have imagined. During the first year, I felt defeated as a professor. I traveled back to Atlanta for the summer and sought advice from my mentor, Dr. Al Baumstark. I returned to Tuskegee that fall semester with new teaching strategies to engage chemistry students in the classroom and to share important life lessons.

Over the years, several of my students have shared reflections about their experiences in my organic chemistry lecture and laboratory courses. One of my students, Dr. Tiffany Mayo Taylor, who is now an Assistant Professor of Dermatology, School of Medicine, at the University of Alabama, Birmingham recently commented on the importance and impact of role models. "Dr. Robinson was confident, approachable, and made herself available to answer questions where the subject often transitioned from chemistry, to career, to life, and back again. She had a significant impact on my academic and professional development," she stated.

My research program at Tuskegee University was in the areas of surface chemistry, natural products, and chemical education. I eventually merged my research areas to address health and educational disparities in the African American communities of Alabama. I secured external funding from NSF, NIH, and ACS (American Chemical Society), which provided critical support for hands-on research experiences for my students and opportunities for them to present their research at national and regional conferences.

Khalifa Jordan, a former student in my research group and now a Technical Project Manager at the U.S. Department of the Treasury, recently shared his thoughts about his undergraduate research experience at Tuskegee University. "Dr. Robinson provided the platform for me to perform natural product research and interact with the larger scientific community through presentations and journal reviews. She mentored me from a place of 'family' and cultivated an environment of STEM exposure," he stated. These reflections from my undergraduate students remind me of the importance of effective mentoring. As an African American woman teaching chemistry at an HBCU, I offered more than scientific and research knowledge to my undergraduate students. I emphasized the significance of grit, self-efficacy, sense of belonging, STEM identity, and resilience.

From Organic Chemist to Consultant

Leaving the academy was the most difficult and important decision I have made in my career. For over 15 years, I was defined by an academic title and my work in the academy. Now, I was moving into a new professional chapter, specifically consulting work. Some years back, I attended an American Council on Education Women's Network Conference and reflected the possibility of being a consulting chemist, but was unsure of exactly what that meant and how I could make a living being a consultant. I talked with my mentors about establishing my own consulting firm. I conducted a SWOT (strengths, weaknesses, opportunities, and threats) analysis of my skills, which I developed while a faculty member at both North Greenville and Tuskegee universities. These skills included grant writing, coaching, program development and evaluation, which are extremely important skills needed for a STEM career.

The next step focused on determining a business name, logo, and mission and vision statement, which were carefully aligned with my purpose driven-work. The business name, PLR Consulting, is the acronym for my initials (Pamela Leggett-Robinson) and the logo is the monogram. My mission and vision statement, "Developing and Delivering Services that Enhance Student and Community Success" aligns with my purpose of broadening participation in STEM and is the guiding principle of every interaction with my clients. The colors of the PLR Consulting logo are green and black, where black represents my race and green represents abundance, growth, and compassion. Shortly thereafter, the website was launched and I began marketing PLR Consulting on social media.

The last step of establishing PLR Consulting truly focused on embracing this new chapter of my career. Furthermore, I spent time surrendering my fear of failure, confronting imposter syndrome, and practicing my elevator pitch for potential clients. Finally, as the Founder and Executive Director of PLR Consulting, I recognize the importance of shifting from a deficit mindset to a growth mindset in order to strategically grow my consulting business and address equity in STEM.

The STEM Butterfly Effect

A butterfly is a flying insect that exhibits beautiful colors on its wings. It undergoes a sequence of changes as it moves through several stages, namely an egg, larva, pupa, and finally an adult (butterfly). The metamorphosis from a caterpillar into a butterfly occurs during the pupa stage. Each stage of the life cycle however, represents growth. As I reflect upon my journey as a chemist, it is truly aligned with the life cycle of a butterfly.

Table 1 illustrates a comparison of my chemistry journey compared to the life cycle of a butterfly. The egg stage of the life cycle represents the early years (undergraduate chemistry education), graduate school at TTU, and doctoral studies at Georgia State University. The pupa and caterpillar stage represents my time spent in the academy as a professor, specifically developing my teaching, leadership, coaching, and program management skills. Most importantly, during this stage I developed a deeper understanding of the significance and impact of simply being a Black woman in chemistry. The butterfly stage is a celebration of learning to fly, embracing the entrepreneurial mindset and achieving new career milestones.

Table 1. A Comparison of the Lifecycle of a Butterfly to the Journey from Organic Chemist to Consultant

Egg	*Pupa and Caterpillar*	*Butterfly*
Early years (Undergraduate chemistry education)	The Professoriate	Entrepreneurial Mindset
Graduate Study at Tennessee Technological University	Leaving the Academy	Celebrating Career Milestones
Doctoral Study at Georgia State University	Development of STEM Consultant Work	Embracing the new Chapter - PLR Consulting

The 'Butterfly Effect' symbolizes the experiences of Black women in STEM. Holistically, it symbolizes the journey of us all. It is the understanding and acceptance that both the process and final outcomes are important. Just as the lifecycle of a butterfly consists of a metamorphosis, we undergo a type of life transformation along our journey. We often focus on the trauma itself instead of how we are being transformed during the process. The journey in STEM as a Black woman is similar to the metamorphosis of the caterpillar. The process is sometimes dark at times, but what lies on the other side is beautiful. When we embrace our personal butterfly effect, we are able to reflect, confront fear, take leaps, become vulnerable, and celebrate the process.

Chapter Summary

My experiences as a chemist have been very diverse. My journey has included both wonderful career milestones and difficult experiences, where I needed to make critical decisions to move forward. My 'Butterfly' experience allowed me to unpack my STEM trauma and focus on my next steps. In life, we all experience personal and professional challenges and we have to successfully navigate these experiences to be successful.

As I look over my journey in chemistry, I am amazed. The woman and chemist that I am today is very different from the undergraduate student I was over 20 years ago. I understand that my perspective and lived experiences within the chemistry culture has an impact on how I engage with others. This chapter highlights the barriers that can contribute to the underrepresentation of Black women in chemistry, while emphasizing the importance of self-reflection, sense of belonging, STEM identity, grit and perseverance.

PLR Consulting is a powerful "chemical reagent" used to transform people and organizations through professional coaching to enhance their skill sets and broaden participation in STEM. Comparatively, dimethyldioxirane (DMDO) is a powerful reagent used in synthetic organic chemistry to prepare organic compounds such as epoxides and sulfoxides. A lesson plan which highlights the chemistry of dimethyldioxirane is provided in the appendix in this chapter. This lesson plan can be implemented in the organic chemistry curriculum when discussing alkene reactions, epoxidation, S_N2 reaction mechanisms and spectroscopy.

Appendix: Sample Teaching Chemistry Assignment

Oxidation of Secondary Alcohols with Dimethyldioxirane

Description

As a doctoral student, Dr. Pamela Leggett-Robinson investigated the chemistry of dimethyldioxirane. In this activity, students will critically review a published article focused on the oxidation of secondary alcohols. Students will use their organic chemistry knowledge to design their own experiments to oxidize a secondary alcohol. The students will work in teams and review the scientific literature to prepare their experiments. Students will use nuclear magnetic resonance (NMR) and infrared spectroscopy (IR) to characterize the products of the reactions.

Learning Goals

There are various learning goals for this assignment.

1) Students will critically read an article published in a high-impact journal.
2) Students will apply their knowledge of organic chemistry concepts to design their own laboratory experiments.
3) Students will characterize the final products from their experiments using NMR and IR spectroscopy.
4) Students will propose an overall reaction mechanism for their final products.
5) Students will prepare a one-page infographic to present their research results to their peers.

Equipment Needs

Internet connection, computer access, text books.

Implementation Notes

The instructor will need to provide a lecture focused on oxidation of secondary alcohols and students will need to access the journal article. An additional resource regarding oxidation by dimethyldioxirane should also be provided to students "Oxidation of secondary alcohols and ethers by dimethyldioxirane" (*10*).

Evaluation

Students can be evaluated by submitting responses to the worksheets for this lesson.

Student Handout

The journal article "The Oxidation of Secondary Alcohols by Dimethyldioxirane: Re-Examination of Kinetic Isotope Effects (*11*)," focuses on the oxidation of secondary alcohols using dimethyldioxirane and their kinetic isotope effects. Dimethyldioxirane (Figure A1) is a versatile, powerful, and selective (chemo-, stereo-, and regio-) oxidizing agent used for a broad scope of synthetic applications. The paper investigates the kinetic isotope effects for the oxidation of a series of deuterated alcohols by dimethyldioxirane to the corresponding ketones in dried acetone at 23 °C. However, for the purposes of this assignment, we will focus on predicting products formed from heteroatom oxidation of a variety of secondary alcohols with dimethyldioxirane. Using the lecture notes and two papers provided, complete the following assignments (Parts 1–3).

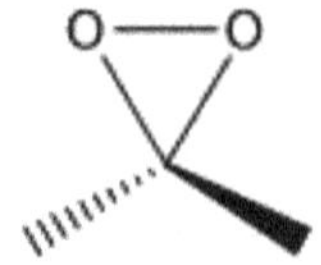

1

Figure A1. Structure of dimethyldioxirane.

Part 1. Oxidation Reactions

Using Scheme A1 and your knowledge of nomenclature and oxidation reactions, complete Table A1.

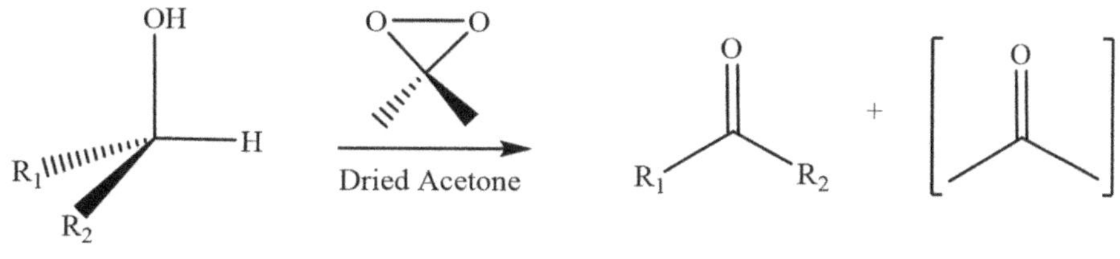

	R_1	R_2		R_1	R_2
1	CH_3	CH_2CH_3	**3**	Ph	CH_3
2	CH_2CH_3	$CH(CH_3)_2$	**4**	$PhCH_2$	CH_2CH_3

Scheme A1. Oxidation of secondary alcohols by dimethyldioxirane.

Table A1. Student Table to Complete for Oxidation Reactions

Structure	*Starting Material Structure*	*Starting Material Name*	*Product Structure*	*Product Name*
1				
2				
3				
4				

Part 2. Nomenclature, Reactions, and Spectroscopy

Use the structures shown in Figures A2 and A3 to answer the questions.

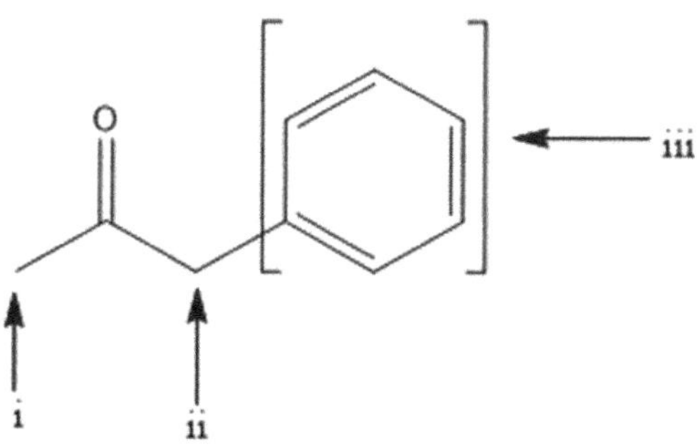

Figure A2. Structure of Ketone 1.

a. Properly name the organic molecule.
b. Which of the protons in the molecule would appear furthest downfield? Indicate the splitting pattern of all of the protons in the molecule.
c. Predict the IR spectrum for the organic molecule shown in Figure A2.

d. What alcohol could be oxidized with dimethyldioxirane to yield the structure shown in Figure A2? (Draw and name the structure).

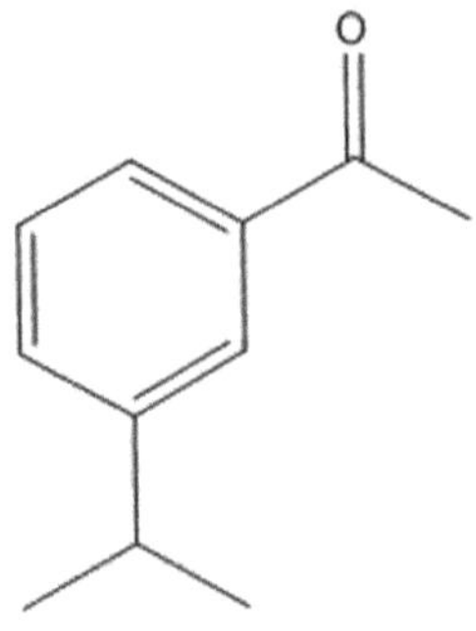

Figure A3. Structure of Ketone 2.

a. Properly name the organic molecule in Figure A3.
b. List the functional groups and corresponding peaks found in the IR spectrum for the compound shown in Figure A3.
c. What alcohol could be oxidized with dimethyldioxirane to yield the structure shown in Figure A3? (Draw and name the alcohol).
d. How many unique protons would you expect to see in the ^{1}H NMR spectra for the structure shown in Figure A3?
e. What splitting patterns would you expect to see in the ^{1}H NMR spectra?

Part 3. Design Your Own Laboratory Experiment

You will work in teams to design your own experiments to oxidize an alcohol. You will review the chemistry literature and prepare an experimental plan for your research project. Your team will characterize the products using NMR and IR spectroscopy. Your team will prepare a creative infographic to present your research results to your peers.

Creative Infographic Resource

https://www.clips.edu.au/infographics/.

References

1. Hurtado, S.; Carter, D. F. Effects of College Transition and Perceptions of the Campus Racial Climate on Latino College Students' Sense of Belonging. *Sociol. Educ.* **1997**, *70*, 324–345.
2. Robinson-Leggett, P. M. Celebrating Transformations Through STEM Storytelling. In *Overcoming Barriers for Women of Color in STEM: Emerging Research and Opportunities*; Robinson-Leggett, P. M.; Villa, B., Eds.; IGI Global: Hershey, PA, 2020, pp 54–81.
3. College Remembers College Professor, Scientist Rubye Torrey. *Syracuse University News*, https://news.syr.edu/blog/2018/10/31/college-remembers-professor-scientist-rubye-torrey-g69/ (accessed July 2, 2020).

4. Hannes, M.; Svatunek, D.; Lumpi, D.; Glöcklhofer, F.; Hametner, C.; Fröhlich, J. Practical and Efficient Large-Scale Preparation of Dimethyldioxirane. *Org. Process Res. Dev.* **2013**, *17*, 313–316.
5. Leggett-Robinson, P. *Epoxidation of Enol Esters and C-H Oxidation of Secondary Alcohols by Dimethyldioxirane*. Ph.D. Thesis, Georgia State University: Atlanta, GA, 2003.
6. Mack, K. M.; Rankins, C.; Woodson, K. From Graduate School to the STEM Workforce: An Entropic Approach to Career Identity Development for STEM Women of Color. *New Directions for Higher Education.* **2013**, 23–34.
7. Mack, K. M.; Johnson, L. R.; Woodson, K. M.; Henkin, A.; Dee, J. R. Empowering Women Faculty in STEM Fields: An Examination of Historically Black Colleges and Universities. *J Women Minor Sci Eng.* **2010**, *16*, 319–341.
8. Mack, K. M.; Rankins, C. M.; Winston, C. E. Black Women Faculty at Historically Black Colleges and Universities: Perspectives for a National Imperative. In *Beyond Stock Stories and Folktales: African Americans' Paths to STEM Fields*; Frierson, H. T., Tate, W. F., Eds.; Diversity in Higher Education; Emerald Group Publishing Limited: Bingley, UK, 2011; Vol. 11, pp 149–164.
9. Davis, J. College in Black and White: Campus Environment and Academic Achievement of African American Males. *J. Negro Educ.* **1994**, *63*, 620–633.
10. Baumstark, A. L.; Kovac, F.; Vasquez, P. C. Oxidation of Secondary Alcohols and Ethers by Dimethyldioxirane. *Can J Chem.* **1999**, *77*, 308–312.
11. Baumstark, A. L.; Vasquez, P. C.; Cunningham, M.; Leggett-Robinson, P. M. The Oxidation of Secondary Alcohols by Dimethyldioxirane: Re-examination of Kinetic Isotope Effects. *Heterocycl. Commun.* **2010**, *16* (4-6), 217–2202010.

Chapter 8

A Multidimensional Career in Chemistry

LaVetta Appleby*

Lawrence Technological University, Department of Natural Sciences, Southfield, Michigan 48075, United States

***Email: lappleby@ltu.edu.**

This chapter focuses on the unconventional road I traveled during my career in chemistry. This chapter will explore my undergraduate and graduate years at Michigan State University, where I earned my master's degree in computational chemistry. I will also discuss my 20-year career as a chemistry educator at Lawrence Technological University in Southfield, Michigan.

Early Years

My journey to become a chemist was by no way a smooth one. I haven't always wanted to be a chemist. In fact, I had very little interest in science. I grew up in a house where my father was an amateur scientist and showed a great deal of interest in everything science. There were always *Popular Science*, *Science* and *Archaeology* magazines around our house. I used to think to myself, how boring! My interest at the time was in photography and fashion. In fact, I planned to be a fashion buyer once I graduated from college.

When I turned 16 years old, I obtained my driver's license and of course I wanted a car. My father made a deal with me, specifically that if I took a chemistry course and earned an "A" he would buy me a new car. I had already completed my requirements for science, one year of biology. Well, of course I took him up on his offer. I enrolled in a chemistry course and earned an "A" but the reward was not the car, it was the fact that chemistry opened up a whole new world for me.

Chemistry gave me the tools to see beyond the surface. Plastic bottles were no longer just plastic bottles but interwoven polymer molecules. Soap was no longer soap but the cleaning action of micelles. I was so excited to learn more, I purchased my first CRC Handbook of Chemistry and Physics as a junior in high school. I took more chemistry and physics courses. I was hooked! I couldn't wait to study chemistry in college.

Undergraduate Years at Michigan State

As a high school senior, I visited several universities and decided that Michigan State University (MSU) was the school for me. What made MSU appealing to me was its' support network for historically underrepresented students. I knew that making a transition from a small high school to a large university, I would need a strong support system. When I visited MSU's campus, the students and faculty were very welcoming and they took the time to speak with me about their programs in detail. I was also introduced to members of several student organizations, including the Black Caucus and the National Organization for the Professional Advancement of Black Chemists and Chemical Engineers (NOBCChE). Although other universities had similar support organizations on campus, the organizations at MSU took the time to meet with me during my campus visit. This was an amazing experience for a high school senior, and I instantly felt that I could reach my goals at this institution. The Black Caucus provided a strong support system to all African American students, especially incoming freshmen. This organization provided a platform for students to freely discuss any issues or concerns that were specific to African American students. It also provided resources to get help in resolving these issues. MSU's NOBCChE Chapter also provided a great deal of academic support. When I became a student at MSU, I joined this organization and I developed friendships with other African American students who were also pursing degrees in chemistry and chemical engineering. I was able to join study groups and obtain peer tutoring. These organizations were instrumental in my acclimation and academic success at MSU.

Upon the advice of my high school advisor, I decided to major in chemical engineering instead of chemistry. She explained that chemical engineering would be a more suitable major for me because of my love for mathematics. After the completion of my freshman year, I obtained a scholarship and an internship from the Dow Corning Corporation in Midland, Michigan. My internship was very interesting and it helped me to discover that I did not want to be a chemical engineer. My project involved the design and the construction of a small scale chemical reactor. This reactor was designed to operate at very high temperatures and pressures. Once the reactor was complete, I spent the rest of the summer testing the reactor. The focus of the project was on the design, construction and testing of the reactor and not on the actual chemical reactions that were taking place within the reactor. I was truly interested in the chemical process and I had very little interest in the design process.

When I returned to MSU, I informed my academic advisor that I wanted to change my major to chemistry and she said that because I was doing so well that I should stay the course. I interned the next summer and my project was very similar to my first project. Finally, my last summer, I worked with a chemist. This project focused on the performance of silicone antifoams in detergents. With this project, I was able to formulate and perform chemical experiments to gain an understanding of how these antifoams work in detergents. I was truly in my element with this project and I enjoyed it immensely. All of my projects were great and extremely thought provoking. I was able to gain experience in working in the chemical industry and these experiences helped me to solidify what career I wanted to pursue.

During my senior year, I was working as a student assistant in a microbiology lab on campus. My duties included maintaining the lab equipment and retrieving journal articles for the lab supervisor. I was approached by the principal investigator (PI) of the research group, Dr. Michele Fluck, and she asked me what my plans were after graduation. I explained that I was planning to accept a job offer from Dow Corning. She asked to see my academic transcript which demonstrated my aptitude for chemistry. She reviewed it and asked if I had considered going to graduate school for chemistry. I met with the department chair of chemistry the next day to discuss graduate school and the following fall semester, I began graduate studies in the department of chemistry.

Graduate Research at Michigan State University

My graduate research efforts focused on using computational chemistry to investigate the nature of lone pairs of electrons of various molecules (*1*). My investigation examined the electron density and the associated molecular electrostatic potential (MEP) for various molecules using ab initio calculations in order to gain greater insight into the nature of the lone pair electrons. My research goals included determining if the lone pair electrons were localized on the central atom as predicted by bonding models such as, the Lewis electron structures and Valence Shell Electron Pair Repulsion model (VSEPR). The molecules studied include ammonia, phosphine, water, hydrogen sulfide, formaldehyde, carbon monoxide, hydrogen fluoride and hydrogen chloride (Figure 1). This series of molecules studied one, two and three lone pair systems, and also examined the effect of atomic size on the lone pair electrons.

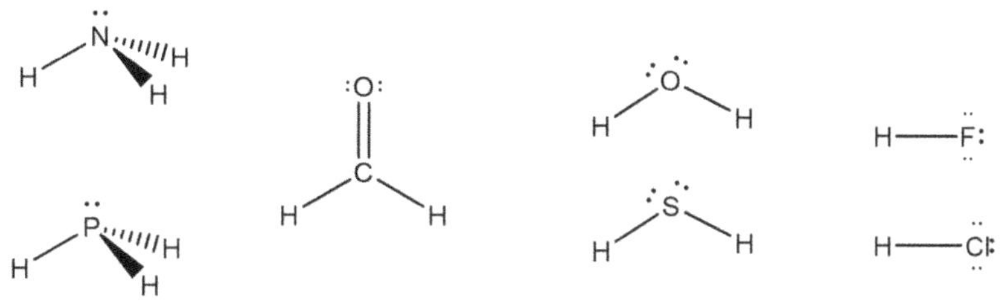

Figure 1. Chemical structures of molecules investigated using computational chemistry.

In 1996, performing these types of calculations were not readily available because the tools needed to perform them were fairly expensive. Today, calculations can be done on laptop computers with inexpensive software. For this reason, computational chemistry has been made readily available not only to graduate students and industry professionals, but also to undergraduate general chemistry students. These type of calculations are very important because students gain a better understanding of Lewis electron structures and the Valence Shell Electron Pair Repulsion (VSEPR) models. These models play an important role in understanding chemical bonding. A lesson plan focused on chemical bonding is provided in the appendix of this chapter. This lesson plan provides students the opportunity to explore and compare chemical bonding models and can be aligned with course content in general chemistry and advanced chemistry courses.

Undergraduate Teaching at Lawrence Technological University

During my graduate studies at MSU, I was a teaching assistant for several general chemistry recitations and laboratory courses. I enjoyed my research, but I also discovered that I enjoyed teaching chemistry to undergraduates. Upon completing my master's degree in chemistry, I decided to find a teaching position. In 1996, I joined the faculty in the Department of Natural Sciences at Lawrence Technological University (LTU), a private institution located in Southfield, Michigan, just outside of Detroit.

The one thing that I have learned in my many years of teaching at LTU, is that students learn by actively participating in the learning process and by applying what they have learned to everyday life. You can introduce science concepts and students will take in the information, and usually, that is the end of the story. They must hear it, write it and apply it. When I first started teaching at LTU, I would lecture for an hour discussing chemistry concepts that were so exciting to me, but when I looked at my students they were half asleep. How could this be when the information I was conveying was so exciting?

When I quickly realized these chemistry concepts were only exciting to me, I had to develop a different approach to teaching. I started including 'conceptual checks' within my lectures. These conceptual checks were problems that were given to the class to solve in the midst of the lecture. I would stop talking and allow students to work together to solve these problems. Students became much more engaged in the classroom. They were more interested in learning these concepts. This was the beginning of getting students actively involved in the learning process. Also, I noticed that their quiz grades improved tremendously! However, their exam grades only slightly increased. However, I realized that I was missing a key component with my classroom teaching. Yes, they were hearing the concepts, writing and applying them by solving conceptual check problems but the concepts did not stick.

Many of my non-chemistry major students had expressed to me that they didn't feel that they should have to take general chemistry because it was not an integral part of their majors. I explained to them that chemistry impacts everyone. If they took a shower this morning, or if they drove a car, then they performed chemical reactions. I realized at that point that the component that was missing was real world applications.

For example, when we study redox reactions, I make sure to discuss the applications of batteries, which are used in our everyday lives. I explained to the students about how the transfer of electrons in a balanced redox reaction, could be harnessed to power a car and many other electronic devices, such as cell phones. We also discussed fuel cells as a viable choice for the replacement of car batteries.

Marvel Studios' Black Panther and Vibranium

Along with real life applications, 'superhero science' concepts featured in popular movies can also be used to engage students. This was a great teaching approach to get students energized and engaged. In discussing Marvel's *Black Panther* with my colleague, Dr. Sibrina Collins, this teaching strategy came to fruition and led to a publication in the *Journal of Chemical Education* (2). We discussed how general chemistry concepts could be applied to the fictitious element vibranium. At that time, I was discussing the properties of elements and their arrangement in the periodic table in my general chemistry class. We were also discussing the Bohr model of an atom and electron configurations. Dr. Collins posed the question, that if vibranium was a real element, where would it be located on the periodic table? I decided to pose this question to my students in my class. I first asked the students how many had seen the movie *Black Panther*. I was not surprised that the majority had already seen the movie and just by asking this question, the students became animated and it provoked a classroom discussion. The students were hooked! I decided to include extra credit questions on their next exam about the fictitious element vibranium. I didn't tell them that the questions were going to be on their exam. The only warning I gave them was that it would be a good idea to go see the movie before the exam, if they had not already done so.

I received a variety of responses to these extra credit questions on the exam. The most important outcome from this exercise was that they began to analyze and discuss the arrangement of elements on the periodic table. They were engaged and wanted to know more about the properties of elements in the periodic table. Relating these concepts to something tangible like popular culture made a huge difference in how students learned the concepts and how they applied them to other problems. I had never seen my students so excited to learn about the periodic table in my 20 years of teaching! My physical chemistry students heard about the discussion on vibranium in my general chemistry class. They asked if they could do a similar activity. This led to an hour-long discussion about vibranium. These students were curious about the possible atomic structure of this fictitious element. They also investigated the possible atomic term symbols of vibranium. I learned that incorporating popular culture into my chemistry curriculum, truly made a positive impact on how my students learned chemistry concepts.

K-12 Teachers and Science Education Outreach

Working with first year students at LTU made me aware of the fact that many students enter college with a lack of the necessary background knowledge needed to succeed in general chemistry and other college level science courses. In 1998, I was asked to teach a physics class for the Detroit Area Pre-College Engineering Program (DAPCEP), which is a 40-year nonprofit organization that partners with universities and K-12 school districts to "connect youth to the best science, technology, engineering, and mathematics (STEM) education experiences in Michigan and beyond (3)."

This class consisted of a basic introduction to Newtonian physics to eighth grade students. I created a curriculum using Vernier Logger Pro (4) sensors to explore motion. One of the motion activities involved students being able to explore velocity and acceleration by walking in front of the motion sensor. They were asked to match the distance vs. time graph on their laptop computer screen. They completed the activity and had a good understanding of velocity and acceleration, but they were not enthusiastic about it. Also, when I asked them about this activity the following week, they were able to recall very little about the activity.

During one class session, I overheard some of the students talking about an episode of the television show Crime Scene Investigation (C.S.I.) (5), which sparked an idea. What if I changed our physics activities by incorporating forensic science concepts? I could keep the same activities but include a background crime scenario. For example, for the motion activity, I added a crime scenario where a precious blue diamond was stolen and the criminal was not detected by the museum's motion detectors. The scenario explained that the only way this was possible was if the criminal was able to recreate the distance vs. time graph. I also added the analysis fiber samples that were left at the crime scene. Students were given samples from each suspect's coat and the one from the crime scene. They analyzed the samples under a dissecting microscope. The fiber analysis helped them make a positive identification of the criminal who stole the diamond. Students got so excited about solving the crime that they did not realize that they were learning about velocity and acceleration! The following week when I asked them about these concepts, they were able to explain them in great detail. I decided at that point, to create crime scenarios for all of our activities and rename the class forensic physics. The director of DAPCEP said that they received positive feedback from students about the course and that more students were now interested in physics.

In 2013, I became the director of the Masters of Science Education (MSE) program for K-12 science teachers. This program allowed teachers to obtain a master's degree in science education. With the implementation of the No Child Left Behind Act of 2002, many K-12 teachers expressed the need to become more competent in teaching science. This program was designed to meet that need. Teachers were not only taught science theory, they were also taught techniques on how to implement science concepts in their classrooms. The program was developed in partnership with the Detroit Zoological Society, Cranbrook Institute of Science, and Aquinas College. The courses incorporated modern hands-on, constructivist-based, inquiry-driven, and project-based learning and teaching methods. The program emphasized active participation and interaction with instructors and classmates which were the same methods of teaching and learning promoted in K-12 science classrooms.

For example, as part of the MSE program, a week-long forensic science for educators' course was offered. This course was offered in partnership with Kathy Marakovits, a Michigan K-12 science educator (6). The course was also taught by myself and other LTU faculty, including Dr. Jeffery Morrissette and Dr. Julie Zwiesler-Vollick. One of the hands-on activities included a body farm. Dead fetal pigs were strategically placed outside and allowed to decompose. They were placed in containers that were opened to atmospheric conditions for several days. Crime scene investigators are often tasked with determining the time of death of a decomposed body. The body farm was a simulation of this scenario. Teachers were asked to determine the time of death of the fetal pigs. The teachers were instructed to remove blowfly larvae from the fetal pigs. Based on the different stages of development of blowfly larvae, teachers were able to determine the time of death of the fetal pig.

This was one of many hands-on activities that the teachers were able to participate in during the week-long course. They were also taught how to set up a crime scene and how to develop a forensics lesson plan for their classrooms. With these tools and resources, teachers were able to walk into their classrooms the next day and teach forensic science. This course equipped teachers with the tools needed to successfully implement forensic science in their classrooms.

In 2015, LTU entered into a partnership with Detroit Public Schools Community District (DPSCD) and created the Blue Devil Scholars Program, which offers a pathway to college for DPSCD students with scholarship support. The objective of this program is to apply a systematic approach to infuse STEM and Design education into the DPSCD curriculum. This program provides hands-on workshops for middle school students, professional development workshops for DPSCD science teachers and dual-enrollment courses for high school students to earn an LTU certificate with 12-15 credit hours towards a college degree.

I have the privilege of working closely with DPSCD teachers and middle school students. Along with my Lawrence Tech colleagues, Dr. James Woodworth and Dr. Fauzia Siddiq, we have offered several hands-on summer science workshops. One of the middle school science workshops was based on the movie *Hidden Figures* (7). This workshop was entitled "Things that Fly," where we explored both the history of flight and the physics of flight (8). Students built paper airplanes, kites, hovercraft boards and several different types of rockets.

In one of the activities, students were given instructions on how to build an airplane glider out of a Styrofoam meat tray. Students explored how the forces of flight affected the performance of their airplane glider. Students were able to test their gliders by measuring the distance and time of flight. The data the students collected were discussed in class and students were asked, based on their data collected, how they could improve the original design. We also discussed how the forces of flight played a role in their airplane gliders performance. Once students had a new design, they modified their airplane gliders and tested them again for distance and time of flight. Students were very excited about their improved designs. This hands-on scientific inquiry-based activity engaged students in the use of the scientific method and it allowed them to develop the critical thinking skills needed to improve their designs. Scientific reasoning is an important tool that students will use throughout their academic careers.

Chapter Summary

My multidimensional career in chemistry has been exhilarating with many twists, turns, ups and downs. Although my journey to become chemist was unconventional, I always landed on the right path. It brings me great joy to watch my students grow and expand their knowledge base in science. Just to see that spark of interest turn into a bright light is why I am doing this important work as a science educator. If you had asked me thirty years ago if I would be where I am today, the answer would have truthfully been absolutely not. My mother instilled in me that a good education is a golden gift to be used for the rest of your life. We all have a gift that we can use to help others achieve their career goals.

Science is both extraordinary and captivating. I am lucky to have the privilege to share my passion for science by working directly with students and K-12 teachers. I have developed creative strategies to teach science, which can make a direct and positive impact on student learning. For example, Marvel Studios' *Black Panther* provided a unique opportunity to engage chemistry students about the organization of the periodic table of elements. As of January 2021, our paper has been downloaded over 19,000 times from the homepage of the *Journal of Chemical Education*. In addition, I incorporated forensic science concepts into my physics course, which proved to be a successful strategy for student engagement. Thus, popular movies and television shows can make a positive impact on students in the classroom.

My multidimensional career as a chemist has allowed me to focus on developing effective teaching practices, which is my passion. You never know where your journey in life will lead you. It is imperative to stay committed to what brings joy to your life. If your journey is similar to my career in chemistry, your passions will lead you to a career where you can make a tremendous impact in your field.

Appendix: Sample Teaching Chemistry Assignment

Bonding Theories: Lewis Structures, Valence Shell Electron Pair Repulsion Theory and Molecular Orbital Theory

Description

In her graduate studies, Professor LaVetta Appleby carried out computational analysis of several molecules, specifically, ammonia, phosphine, water, hydrogen sulfide, hydrogen fluoride and hydrogen chloride. In this activity, students will sketch Lewis structures of each molecule, determine the point groups and explain the bonding of these molecules using Valence Shell Electron Pair Repulsion Theory (VSEPR) and Molecular Orbital (MO) Theory.

Learning Goals

There are various learning goals for this assignment:

1. Sketch the Lewis structures for each molecule.
2. Determine the point group for each molecule.
3. Explain the molecular structures using bonding theories.
4. Develop a creative infographic to explain the key differences between Lewis structures, VSEPR and MO theory.

Equipment Needs

Internet connection and computer access.

Implementation Notes

The instructor will need to provide students with a lesson on determining point groups for molecules along with a flow chart. Students will also need access to several character tables. Another creative strategy educators can use when teaching this topic is focused on the generation of ligand group orbitals (LGOs). (Johnson, A.R.; Nataro, C. Teaching Molecular Orbital Theory Better. In *Advances in Teaching Inorganic Chemistry Volume 1: Classroom Innovations and Faculty Development*; Jones, R.M., E.d.; ACS Symposium Series 1370; American Chemical Society: Washington, DC, 2020, pp 47-63).

Evaluation

Students can be evaluated by submitting responses to the worksheet for this lesson.

Student Handout

The Lewis Electron Dot model provides the arrangement of electrons surrounding the central atom of a molecule and the Valance Shell Electron Pair Repulsion (VSEPR) theory allows us to predict the molecular geometries. The VSEPR model generally predicts the correct shape of a molecule, however, it does not provide an explanation of chemical bonding. In this activity you will investigate the Lewis Structure and VSEPR theories to determine the electron arrangement and the molecular geometry for the following molecules: NH_3, PH_3. H_2O, H_2S, HF and HCl. You will also use Group theory and Molecular Orbital theory to explain the chemical bonding.

Part 1 Lewis Structures and VSEPR Theory

1. Determine the total valence electrons for each molecule and sketch the Lewis Structures for the following molecules: NH_3, PH_3, H_2O, H_2S, HF and HCl.
2. The steric number of a molecule is the number of atoms bonded to the central atom of a molecule (σ sigma bond) plus the number of lone pairs on the central atom. It is often used in the VSEPR theory in order to determine the molecular geometry. Determine the steric number for each molecule in step1 and use the VSEPR table below to determine the molecular shape.

Table A1. VSEPR Table

Steric Number	*Hybridization*	*Electron Group Geometry*	*Ligand Number*	*Number of Lone Pairs*	*Molecular Shape*	*VSEPR Bond Angles*
2	sp	Linear	2	0	Linear	180°
3	sp^2	Trigonal Planar	3	0	Trigonal Planar	120°
			2	1	Bent	120°
4	sp^3	Tetrahedral	4	0	Tetrahedral	109.5°
			3	1	Trigonal Pyramidal	109.5°
			2	2	Bent	109.5°

1. Group theory is a great tool that can be used to explain chemical bonding in molecules. You will use molecular symmetry to determine which atomic orbitals contribute to molecular orbitals to form chemical bonds. Use the flowchart provided by the instructor to determine the point group for the H_2O molecule.
2. The electron configuration for each hydrogen atom in the H_2O molecule is $1s^1$. The two 1s atomic orbitals are available for mixing to form molecular orbitals (chemical bonds) with the oxygen atomic orbitals. The two 1s atomic orbitals will be treated as a linear combination of atomic orbitals. This can be done by determining the way in which the hydrogen atoms can combine, in-phase or out-of-phase. Refer to Figure A1 at the end of this list. There are two linear combinations of the two hydrogen atomic orbitals. These are called symmetry adapted linear combinations (SALCs) or ligand group orbitals (LGOs).
3. Assign x, y, z coordinates (z axis is principal axis) to the SALCs.
4. Use the symmetry operations for the point group determined in step 1, to determine the Mulliken symmetry symbols. For the H_2O molecule, the symmetry operations for the C_2V point group are E, C_2, $\sigma_{V(XZ)}$ and $\sigma_{V(YZ)}$. Sketch the hydrogen SALCs as they undergo each symmetry operation. If the orbitals stay the same, you will assign a character of 1, if they change, you will assign a character of -1.
5. You have just determined the irreducible representations which correspond to the SALCs. Use the character table for your point group determined in step 1 to determine the Mulliken symbols for the in-phase SALC and the out-of-phase SALC.
6. The electron configuration for oxygen is $1s^2 2s^2 2p^4$. The 2s and 2p atomic orbitals are available for mixing with the hydrogen SALCs. Use the character table for the point group you determined in step 1 to determine the Mulliken symmetry symbols for the 2s, $2p_x$, $2p_y$ and $2p_z$ atomic orbitals.
7. To form the molecular orbitals, the oxygen atomic orbitals must have the same symmetry as the hydrogen SALCs. Combine the atomic orbitals from the oxygen atom with the hydrogen SALCs of same symmetry and similar energy. Sketch the MO diagram including a pictorial representation of the MOs formed.
8. Repeat Steps 1-7 for the following molecules: H_2S, NH_3, PH_3, HF and HCl. Do you need to determine SALCs for the hydrogen atoms in the HF and HCl molecules? Explain.
9. Choose one molecule and create an infographic to explain the key differences between the Lewis structure, VSEPR and MO Theories.

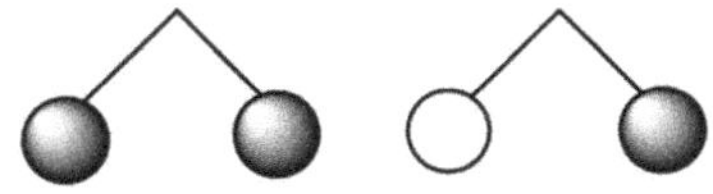

In-phase Out-of-phase

Figure A1. SALCs of H atoms in-phase and out-of-phase in-phase.

Questions

1. Does the molecular orbital diagram agree with the Lewis Structure determined in Part 1? Explain.
2. Does the molecular orbital diagram agree with the geometry predicted by VSEPR in Part 1? Explain.

References

1. Appleby, L. M.S. Thesis, Michigan State University, Lansing, MI, 1996.
2. Collins, S. N.; Appleby, L. Black Panther, Vibranium and the Periodic Table. *Journal of Chemical Education* **2018**, *95* (7), 1243–1244.
3. *Detroit Area Pre-College Engineering Program*, https://www.dapcep.org/ (accessed Nov 15, 2020).
4. *Vernier Logger Pro*, https://www.vernier.com/product/logger-pro-3/ (accessed Nov 15, 2020).
5. *Crime Scene Investigation (C.S.I.)*, https://www.cbs.com/shows/csi/ (accessed Nov 15, 2020).
6. Siegal, B.; Mirakovits, K. *Forensic Science The Basics*; CRC Press: Boca Raton, FL, 2015.
7. Collins, S. N.; Appleby, L.; Kustarz, M.; Zhu, G-C.; Phillips, T.; Brown, T.; Markiewicz Curricular Transformation and Hidden Figures: A STEAM Middle School Camp for DPS Students. *Michigan Science Teachers Association Newsletter* **2017**, *69* (4), 16–19.
8. Mason, K.; Rodenberg, S.; Kravik, R.; Polk, J.; Freit, B.; Schauf, S.; Brewer, B. If you build it, will it fly? *Science Scope* **2014**, *37* (8), 16–20.

Chapter 9

Biochemistry and Leadership in Academic Affairs: When Inspiration Touches the Heart

Angela W. Peters*

Albany State University, Albany, Georgia 31763, United States

***Email: angela.peters@asurams.edu.**

This chapter focuses on the intellectual achievements and career of Dr. Angela W. Peters in chemistry as a student, college professor, department chair and provost at HBCUs (Historically Black Colleges and Universities). She describes effective mentoring and offers tools and tips on leadership in academic affairs.

The Road Less Traveled

Growing up in a family of educators and servant leaders paved the way for me and my siblings from Orangeburg, SC. Expectations were high in my household and education was our vaccine for poverty. Born in the mid-1960's to William Wade and Roberta Rhue Waymer, I was the oldest child (Angela Yvette Waymer) with a brother, Wade Anthony Waymer and a sister, Brandi Janette Waymer (Berry). I learned at an early age the value of a college education and the alignment of academic performance to career, financial stability and independence. Both of my parents had master's degrees in education and my father furthered his education by receiving his doctorate in Higher Education Leadership from Florida State University. My father was a lover of exploration, traveling and good food. We lived in Albany, GA and Tallahassee, FL prior to settling down in South Carolina.

My parents are products of an HBCU. My mother worked her way up the career ladder in the local school district from being a music teacher, guidance counselor and eventually assistant principal. My father also worked his way up the career ladder from a middle school teacher in the local school district to an executive administrator at the local university. As with so many children, my parents were my first teachers of life. I contribute my understanding of family-centered values and my ability to multi-task to my mother. My father taught me how to be a well-informed and respected businesswoman. He also taught me the value of hard work and that first-impressions are tremendously important in life. My father passed away in 2006. That is when I realized the profound influence he had on my life as an educator and parent. He and my mother were an unstoppable team. I am who I am because of them. Nobody rises to low expectations!

In the 10th grade, I had challenges with a chemistry course. Making the connections between the periodic table, measurements and stoichiometry were difficult for me. For a student to love chemistry, you should enjoy reading and mathematics. My father found me a tutor. She was a college student, chemistry major and a member of the student affiliate chapter of the American Chemical Society (ACS). My "F" grade eventually improved to a "C." Subsequently, I completed the course and earned a "B" as a final grade. The proverbial light-switch was turned on and my curiosity in chemistry was peaked, all because of a teacher.

Navigating to and through College

When I decided to attend Hampton University, an HBCU located in Hampton VA, it was because the university had a stellar reputation, one of my favorite aunts and role model, Gloria Waymer Dean, graduated from Hampton, and many of my high school friends attended the institution. When it was time to decide on my major, my father suggested chemical engineering. I had no idea what chemical engineers did for a living. We did not have the Internet or Siri in the early 1980's. However, he did tell me that it involved chemistry, math, and physics and it would yield me a job after graduation with lots of income. Hampton University did not have a chemical engineering program, so the next best thing for me was chemistry.

How can you be prepared for STEM (science, technology, engineering and mathematics) in college? There must be preparation on the front end, and it takes much more than K-12 classes. The collaboration between local colleges and school districts is paramount to developing a successful college student. I am a huge proponent of summer camps, summer enrichment programs, and summer bridge programs. When I entered college, I was familiar with college life because of my parents and the field experiences I took with my father and his students. I was raised on a college campus and I knew about the academic structure and senior leaders. What I was not prepared for was the immersion in college level courses and the importance of having a good relationship with your advisor and your Instructor.

The women mentors at Hampton were instrumental in molding me and guiding me through the curriculum. Dr. Willie Darby and Dr. Nandita Banerjee were my primary advisors. They did more than help me with my schedule. They allowed me to sit in their office and eat my lunch, while waiting for my next class. They also knew me by name and they inquired about my performance in classes. Our mentor and mentee relationships grew over my undergraduate years and they recommended me for a position as a Laboratory Technician at Sun Refining and Marketing in Marcus Hook, PA upon receiving my bachelor's degree in chemistry.

Navigating the Path to Industry and Master's Degree

My parents were so very proud of me when I took the position at Sun Oil Refinery in Marcus Hook, PA in the mid 1980's. I held a low ranking position as a Laboratory Technician, but these technician roles are critically important for the chemical industry. My uniform consisted of overalls, I reeked of oil, and I worked third shift in the distillation lab. In addition, I was required to lift almost 60 pounds of heavy pots and I weighed about 116 pounds at the time. Nevertheless, there were good people who worked with me at Sun Oil. When a position opened in my laboratory, I contacted one of my chemistry colleagues from Hampton, Cheri Hawkins from Philadelphia about the job opportunity. Cheri and I worked together for 4 years at Sun Oil. I eventually was promoted from third shift to first shift. The pay was low and my rent was high, but the experience was invaluable. Working in industry gives you an opportunity to apply your classroom knowledge to real chemistry applications, which are important for society. This is experiential learning at its best. At Hampton, I volunteered to work in the organic chemistry lab on Saturdays. This was the beginning of my undergraduate research experience. It paid off greatly and I was able to apply what I learned in the organic lab towards hands-on distillation of large-scale batches of petroleum at Sun Oil Refinery.

I soon realized that in order to increase my salary at Sun Oil and being promoted to the level of a Chemist, I would need an advanced degree. Thus, I took advantage of the employee education assistance program and enrolled in graduate courses at the University of Delaware. This was very difficult with my work schedule and the courses were held in the evenings. However, I believe it is important to never burn bridges and stay in communication with your mentors. Dr. Darby, my college advisor, contacted me about applying for a fellowship to receive my graduate degree in chemistry from Hampton. The university received a grant from National Institutes of Health (NIH) that would pay my tuition and provide a monthly stipend.

The time that I spent in industry strengthened my research acumen for graduate school. Upon my return to Hampton for graduate school my undergraduate mentors were still at the university, namely Dr. Darby, Dr. Banerjee and my new thesis advisor Dr. Mary Bechtold. Many of the professors at Hampton University, including the department chair, Dr. Isai Urasa, served as role models to prepare students to be leaders. The research project for my M.S. degree focused on the enzyme characterization and purification of catalase from oysters. I completed my master's degree in two years and knew I would further my education and obtain a doctorate. At every juncture in my life, I had mentors and people whose shoulders I stood on. I wanted to give back to so many people who helped me along my journey. This is what motivated me and was my driving force to reach for greater heights in my career.

Navigating the Path to Doctoral Degree and Administration

The University of South Carolina (USC) offered me a teaching assistantship with paid tuition and a monthly stipend. My advisors were Dr. Sandra Berger and Dr. Bruce Dunlap. It was a great arrangement and the best of both worlds. Dr. Berger was a Professor in Basic Pharmaceutical Sciences and Dr. Dunlap was a Professor in Biochemistry. During my four and a half years as a doctoral student, I worked just as hard as I did as a shift worker at Sun Oil. My days were on the average 12 hours during the week and 8 hours on the weekends. My research focused on the effects of ligand binding at the active site of human thymidylate synthase (*1*, *2*). Many students often transition directly from undergraduate to doctoral programs. This was clearly not my path. I needed to understand chemistry more. The master's degree gives you an opportunity to take advanced chemistry courses. This truly helped guide me as a doctoral student. Dr. Berger and Dr. Dunlap were pivotal in helping me develop as a researcher and scholar.

While at USC, I took advantage of every opportunity to learn more about my research and the world of science around me, particularly professional development opportunities. There were very few underrepresented minority undergraduate students at the university. Thus, as a graduate student, when it was appropriate, I was asked to serve as a mentor for undergraduate STEM majors. Serving as a mentor for the National Science Foundation funded South Carolina Louis Stokes Alliance for Minority Participation Program (SC-AMP), was exhilarating and purposeful. I was more than a teaching assistant and a tutor. I could actually help guide students to reach their full potential both in life and STEM.

Upon receiving my doctorate, I was approached by the Dean of the College of Engineering at USC to serve as the Executive Director for SC-AMP. USC was the lead institution and the Dean of Engineering was the Project Investigator. I served in this capacity for three years and led a consortium of 8 institutions and more than 9,000 undergraduate underrepresented minority students. The goal of SC-AMP was to significantly increase the number of underrepresented minority students who earn degrees in STEM. As the Executive Director for the state of South Carolina, I took advantage of trainings and certifications relative to leadership and management. I learned to manage a multi-million dollar federal grant and how to work effectively with a team of educators from presidents, to academic leaders, to faculty and students. Dr. Judith Salley-Guydon, Professor of Biology at South Carolina State University, an HBCU set the tone for undergraduate research and peer-mentoring in STEM. She was an amazing mentor for me, while I led SC-AMP. The success of SC-AMP has resulted in thousands of underrepresented minority students from South Carolina institutions graduating with a STEM degree. Moreover, the program is still thriving after 25 years of success and Dr. Salley-Guydon and SCSU ended up taking the lead on the grant following my departure from USC.

Navigating the Path to Academia

Dr. Vermelle Johnson, Vice President of Academic Affairs at Claflin University, an HBCU and Dr. Verlie Tisdale, Dean of Natural Sciences, Mathematics and Computer Science recruited me from USC to serve as Chair of Chemistry and Associate Professor. Our friendships developed and we found our passion in outreach to the K-12 community. For more than 15 years, we led a statewide STEM mentoring program for girls that was funded by the Sunshine Ladies Foundation. Through the GEMSS (Girls Emulating Maturity Strength and Scholarship) Program, we mentored and supported hundreds of girls in the STEM pipeline from elementary school through college. In 2011, the South Carolina section of the ACS received a ChemLuminary Award from the ACS Women Chemists Committee for outstanding outreach to girls in elementary education based of the GEMSS Program at Claflin. In addition, Dr. Tisdale and I often dreamed about starting a STEM Charter School. Little did we know the new superintendent (Dr. Cynthia Wilson) of the Orangeburg Consolidated School District 5 would seek our expertise in establishing the first High School for Health Professions in the state of South Carolina. Dr. Cynthia Wilson is a visionary and another mentor of mine. I am forever grateful to Drs. Johnson, Tisdale, Salley-Guydon, and Wilson for their friendship, wisdom and innovative ideas. They are exemplary leaders, who I respect and emulate with pride.

I was successfully promoted at Claflin University over a 19-year period as faculty, chair, associate professor, professor and vice provost of academic programs. The late Dr. Howard Hill was another Vice President of Academic Affairs and he was pivotal in preparing me for early tenure (3rd year) as well as being promoted to full professor in my 4th year. I hit the ground running with writing grants (federal and private), developing partnerships with industry, government labs and school districts, and garnering approval of the chemistry department by ACS. One of my most pivotal strategies as chair of chemistry was hiring the right people to teach our students. The team that we put together was one of the best and included experts in every area of chemistry including two exceptional African American women chemists, Dr. Sibrina Collins and Dr. K'Shaun Sanders. We put as much time and energy into recruiting faculty as we did recruiting undergraduate chemistry majors. Our results were astonishing. We increased our chemistry majors from 25 to 80 in a short period. Following this was an increase in undergraduate research placements, graduate and professional school placements, student-faculty peer-reviewed publications and external grants received. Based on the success I had in the classroom, with research and service, the State of South Carolina bestowed upon me the honor of Governors Professor of the Year in 2005. Likewise, in 2004 I received an honor from the White House Initiative on HBCUs for the National Millennium Award for Excellence in Teaching. Student success in chemistry and biochemistry at Claflin University was infectious not only for students, faculty and staff but for the community. Our strategies for capacity building and strengthening the academic enterprise was based on collaborative learning, integrative pedagogy and fostering a growth mindset where students understand how best they learn (*3*, *4*). Advancing diversity in the chemical sciences and promoting inclusiveness within the community garnered me the honor of receiving the 2010 Stanley C. Israel Regional Award for Diversity and the 2017 E. Ann Nalley Southeastern Regional Award for Volunteer Services from the ACS.

As Vice Provost, I was able to develop policies and procedures that brought about systemic change for the university. Being a viable part of the President's executive leadership team was a tremendous revelation and opportunity. I learned how the institution operated from enrollment management to student success to financial aid to institutional effectiveness, just to name a few. President Henry N. Tisdale and Provost Karl Wright were my direct supervisors and they taught me how to be an efficient and effective leader, how to think and act strategically, and never losing sight of the bigger picture (mission and vision of the university). The purpose-driven work that I was involved in as Vice Provost at Claflin was transformative and student-centered. Through the Bill and Melinda Gates HBCU Transformation Initiative and the UNCF Career Pathways Initiative funded by the Lilly Endowment, we were able to increase retention, increase graduation rates, enhance the first-year experience to include learning communities, infuse research into teaching and learning (*5*) and incorporate high-impact practices to promote student achievement in STEM (*6*).

Currently as Provost and Vice President of Academic Affairs at Albany State University (ASU), I continue my life-long work of mentoring, teaching, molding, guiding, relationship building and influencing. My lineage of women mentors continues at ASU with a woman President, Marion Ross Fedrick and Dr. Wendy Wilson as chief of staff. With the remainder of the President's cabinet members predominantly women, we are committed to student success and the mission of the institution. Similar to all the mentors in my life who played a major part in my development, my ASU family continues to inspire me.

The opportunity to hold executive leadership positions at two stellar HBCUs to impact critical academic policy and curriculum changes for students, faculty and staff is a chance of a lifetime. Claflin University has achieved recognition by *U.S. News & World Report* as one of America's best liberal arts colleges. Albany State University is ranked nationally as one of the most affordable colleges for distance learning, one of the top producers of nurses for the state and the largest HBCU in the state of Georgia. Both institutions are powerhouses within their region and greatly fuel economic growth and drive innovation.

Mentoring African American Women Chemists

My chemistry journey highlights the importance of agency in building a network of mentors, giving back to my community and leading with courage and humility. There are many exceptional women scientists I have met and mentored throughout my life. Just like me, they were faced with challenges throughout their career which required them to reflect about themselves and their dreams and aspirations. Many themes that resonated amongst these women include the following elements and serves as the foundation of the experiences that I have highlighted:

- Life-work balance
- Promotion and opportunities for advancement
- Service and contributions to the university and the community
- Appreciated and respected for expertise within the chemistry discipline
- Gender and wage gap

My list of "Giants" includes women who have served as my role models, mentors, mentees, teachers, colleagues, sisters and friends. These women come from diverse ethnic backgrounds and religions; varying ages and different socioeconomic status. Three of these women allowed me to hear their story and their journey from the chemistry classroom to leadership positions in their respective fields. The contributions of these women to chemistry are more than publications and media profiles; they are the silent warriors who have worked in the trenches as servant leaders.

The stories of three African American chemists, specifically Letitia Ling, Dr. K'Shaun Sanders and Dr. Sibrina Collins are shared below as informal interviews. My hope is that my story and their narratives will encourage historically underrepresented women to learn to love science as we do, and to stick with it!

Leticia Ling, Major: Chemistry

Occupation: Product Assurance Inspector, Level 4 at Westinghouse Electric Company

Tell me about growing up and how you fell in love with science.

I grew up in a small rural town called Garnett, South Carolina. I am the fourth child (and a twin) of six children born to the parents of Rev. James and Florence Ling. After high school, I worked to help take care of my family. I attended college at the age of 27 and my first love was art. I fell in love with science after my very first chemistry class.

Tell me about your journey to and through college to complete your degree.

My love for art led me to Denmark Technical College to pursue an Associate's Degree. Instead of enrolling into art classes, something led me to science. That first science class at Denmark Technical College sparked an interest, I never knew existed. After two years at Denmark Technical College, I transferred to Claflin University eager to work toward my B.S. in Chemistry. Claflin University opened doors to incredible opportunities. Being able to participate in undergraduate research provided me the knowledge and skills that I currently use as a Laboratory Technician.

Talk about your mentors and who inspired you to remain in the sciences.

Mrs. Margaree Bonnette (guidance counselor) helped me transition to Denmark Technical College, after such a long break from high school. Dr. Angela Peters (chair of chemistry) was the leader of my strong support system at Claflin University. She encouraged me to continue through to the end.

What are your career goals?

My career goal is to become a chemist at Westinghouse. In order for this to happen, I have to obtain my graduate degree. The company will pay for this and I look forward to being in a position where I can be promoted.

Dr. K'Shaun Sanders, Major: Chemistry

Occupation: Assistant Administrator, Westwood High School (Former Instructor and Chemistry Laboratory Manager at Claflin University, Certified Chemistry High-School Teacher)

Tell me about growing up and how you fell in love with science.

I grew up in the very small community of Cross, South Carolina. My mother had me when she was a high school senior. She and my dad were married when I was two years old. Family surrounded me growing up and one of my cousins (Cindy) had a tremendous influence on me. We shared the same birthday, she was smart, outgoing, and I was just enamored with everything she did. Cindy was 6 years older than me and everything I wanted to be. She went to an HBCU, majored in chemistry, pledged Delta Sigma Theta, and had a great job as a chemist upon graduation. Seeing how successful she was encouraged me to strive for excellence and to pursue a science degree. In addition to my cousin, my Godmother attended the University of South Carolina and earned a chemistry degree. When I became a senior in high school, I was heavily influenced by intelligent Black women who earned STEM degrees. My cousin, Godmother, and several women teachers had given me the confidence and example to believe that I could achieve anything in life. My mother and father had limited educational experiences, but they gave me a stable life growing up and gently encouraged me to seriously consider science, specifically medicine as a career. As I reflect on my childhood, I have other family members that loved science too. For example, my uncle was a math teacher, and I had another uncle who was also a chemist. All of them were important and successful examples of STEM professionals for me.

Tell me about your journey to and through college to complete your degree.

Growing up in rural South Carolina the necessity of pursuing a college degree was instilled in me at an early age. The expectation was that my brother and I would excel in elementary, middle, and high school and then matriculate through college. While attending a summer science program at Clemson University before my senior year of high school, I met a young woman from Ridgeland, SC that changed my path. Until then, I had plans on attending the University of South Carolina with a major in chemistry to become either a teacher or a pharmacist. I am embarrassed to admit that I had never heard of Spelman College until the summer of 1992. After meeting her and doing some research on the prestigious institution, I was committed to attending Spelman University, if I could get in. After returning home later that summer, I talked to my mom about attending Spelman and the experiences I could have as a Spelman student. My mom was in full support. I applied for early decision in the fall of 1992 and was accepted to my dream college.

I started Spelman College in the summer of 1993 with a small cohort of women from around the country, who were pursuing a degree in the sciences. I took full advantage of my early start at Spelman College before the fall semester began. During my four years at Spelman College, I met lifelong friends and immersed myself into my chemistry major, which proved not to be nearly as easy as my high school career. There were several times during my college career when I contemplated changing majors or changing schools because it was hard. Thankfully, I did neither. I joined study groups and formed mentoring relationships with my professors. I was able to participate in research projects and other experiences that further cemented my career choice of becoming a teacher. I had always known deep down that I wanted to be a teacher, primarily because of two phenomenal women science teachers, Mrs. James and Mrs. Bryant. I knew I wanted to be a science teacher, which is why I majored in chemistry.

Talk about your mentors and who inspired you to remain in the sciences.

The women in my life have definitely played a major role in inspiring me to remain in the sciences. Even though my mom only had a high school diploma, her relentless support and belief in me cannot be overstated as having a direct result in remaining a chemistry major during some rough courses at Spelman. Besides my mother, I have been fortunate to meet some amazing scientists along the way. Professor Cornelia Gillyard, a chemistry professor at Spelman College was very impactful as she gave me my first research opportunity. Furthermore, Professor Gillyard traveled with me to my very first research conference in New Orleans, LA and helped me to earn a chemistry scholarship while I was a student at Spelman College. She remained a mentor throughout my education at Spelman College. I also had great mentors during my career. Specifically, Dr. Tisdale and Dr. Peters at Claflin University were a positive influence in my decision to not only remain in the sciences, but to continue in education. These two women were the heart and soul of the science departments of Claflin University. They provided me with my first academic position in higher education and provided me with experiences, advice, and encouragement to continue on my path in science education. For each of these women, I am eternally grateful and I pay it forward with not only my daughter, but also my other family members and students that I have met along the way.

What are your career goals?

I earned my doctorate in Curriculum and Instruction in 2012 from the University of South Carolina. Go Gamecocks! As far as my career goals, I want to continue on my path of developing programs and opportunities to support high school students in their journey to achieve their goals. I may continue to work as an administrator, or pursue another opportunity in higher education, which will allow me to continue my career passions and walk in my purpose.

Dr. Sibrina Collins, Major: Inorganic Chemistry

Occupation: Executive Director of the Marburger STEM Center, Lawrence Technological University

Tell me about growing up and how you fell in love with science.

Growing up in my hometown of Detroit, Michigan, I didn't really consider a career as a scientist. However, I actually fell in love with chemistry as a community college student. I enrolled in a chemistry course for non-science majors at Highland Park Community College and believed the course was easy. I actually really enjoyed stoichiometry and dimensional analysis.

Tell me about your journey to and through college to complete your degree.

I earned my associate's degree in science from Highland Park Community College in 1990. I later enrolled at Wayne State University (WSU) in Detroit, MI and earned my B.A. in chemistry with the support of the National Institutes of Health's Minority Access to Research Careers (MARC) program. As a MARC Scholar, I received a wonderful hands-on research experience in the laboratory of Professor Regina Zibuck, an organic chemist at WSU. My research project focused on the synthesis and characterization of Wieland Miescher Ketone, which is useful for natural product chemistry.

Talk about your mentors and who inspired you to remain in the sciences.

Professor Zibuck actually inspired me to earn my doctorate and encouraged me to pursue graduate studies in the Department of Chemistry at The Ohio State University (Columbus, Ohio.) Furthermore, Dr. Keith Williams and Professor Joseph Francisco also served as mentors to me at Wayne State. However, after graduating from WSU in 1994, I learned about chemistry pioneers such as Dr. Marie Maynard Daly, Dr. Saint Elmo Brady, Dr. Samuel P. Massie, Dr. Bettye Washington Greene and so many others.

Earning my PhD in chemistry has provided me with amazing career opportunities in academia, including my first faculty appointment in the Department of Chemistry at Claflin University, where Dr. Angela Peters was my department chair. Dr. Peters was an amazing leader and mentor to me, as I navigated my new role as an assistant professor of chemistry at Claflin University. It is absolutely critical for women in leadership positions to mentor and support other women for success.

My late grandmother, Geraldine Gray is also an important inspiration for my career. My grandmother was an educator in Detroit and would often deliver presentations in the community focused on the important contributions of African American scientists and inventors. I am truly proud that I am continuing my grandmother's legacy to address equity in STEM.

What are your career goals?

My current role is the founding executive director of the Marburger STEM Center on the campus of Lawrence Technological University. My role is really to promote STEM education on campus and beyond. Throughout my chemistry career, I have focused on encouraging the next generation to pursue a career pathway in STEM. I openly share my career challenges in STEM with students and emphasize the importance of finding your own happiness and pursing your passions in your career.

Outlook on STEM Representation

Despite progress that has been made over many decades, minorities, women, and economically disadvantaged students remain underrepresented in STEM (*7*). According to the National Science Foundation (2018), women make up half of the total U.S. college-educated workforce, but only 28% of the science and engineering workforce (*8*). Hispanics, Blacks, and American Indians/Alaska Natives make up a smaller share of the science and engineering workforce (11%).

The persistence of gendered paths in career choices has recently been reflected in the current Global Gender Gap Report of the World Economic Forum (WEF), which states that on average men are underrepresented in the fields of education, health and welfare whereas women are underrepresented in the STEM fields (*9*). Research shows that approximately half of the African American and Native American freshmen entering STEM majors drop out or switch majors and two-thirds of Hispanic students do not complete their degrees (*10*). Realizing the importance of the momentum year and the sophomore year at all institutions where I served, we implemented a university-wide initiative focused on student success, specifically targeting the freshman and sophomore year experience. The sense of belonging, inclusion and being a valuable, respected and contributing member of the institution captured me and motivated me for three decades in higher education to remain steadfast and stay the course. It is absolutely imperative to emphasize the importance of diversity, equity and inclusion to all stakeholders within an institution.

My success in chemistry over the past 38 years was not solely the acquisition of knowledge in my discipline, but more about gaining leadership skills, building a network of mentors, and building meaningful and positive relationships over time. In my opinion, professional development and leadership development are critical to the success of an effective leader. Professional development helps strengthen and maintain competence in your profession so that you are up to date with skills and knowledge in the field. However, leadership development builds and expands your capacity to perform in leadership roles and focuses on the practical expertise that you need to develop yourself while being efficient in your role within the organization. I was recommended for several leadership development opportunities by the presidents I served, and these opportunities were critical to my success in higher education. Particularly the following: Harvard University Management Development Program; UNCF & American Express Corporation, 360 Leadership and Coaching Project; UNCF Women's Leadership Development; and National Businesswomen's Leadership Association, Leadership Development and Management. Furthermore, my presidents allowed me to be creative and to serve as a leader for not only the institution but the community and the region, particularly as I served as General Chair of the Southeastern Regional Meeting of the American Chemical Society (SERMACS); Secretary of SERMACS; President of the SC Section of the ACS; and Board Chair for the High School for Health Professions.

At every juncture of my career, I was blessed to have a mentor who did more than guide me to the finish line. They were courageous enough to have those crucial conversations with me when I needed constructive criticism. My leadership journey has been paved with a number of exceptional leadership examples. Moreover, the experiences garnered throughout my career have each been established on the heels of "Giants" in our profession. As I contribute to this great profession, my hope is that the phrase "Each One, Teach One" rings resoundingly as I influence those watching me.

Chapter Summary

The overarching theme throughout this chapter is the importance of women role models and mentors in STEM. In order to attract and retain women scientists, we must utilize effective strategies that have proven to be impactful such as mentoring, professional development activities and providing role models for support and encouragement. The mentoring arrangement can be informal or formal such as one-on-one peer mentorship arrangements, or even programs that match new faculty with alumni and local professionals. Unfortunately, girls and women who venture into STEM often find themselves pushed out of the field because of feelings of isolation or not feeling confident in their own skillsets. This chapter provides experiences and evidence of how to attract and retain more women in STEM while preparing them for leadership roles within the discipline. Some students inherently love science at an early age and some students find their love for science while in high school or even when they enter college. I am confident that conversations surrounding outreach, mentorship, professional development, leadership development and experiential learning will enhance the overall experience of women students in STEM, and provide a launchpad for their promotion throughout the academy, industry, research laboratory and in the K-12 classroom. Lastly, I have included an engaging biochemistry lesson plan in the chapter appendix, which will build upon the student's knowledge of protein structure and protein-binding interactions as they evaluate structures of human thymidylate synthase using the Protein Data Bank (PDB). This lesson is based on one of my first peer-reviewed publications as a doctoral candidate (2).

Appendix

Sample Teaching Chemistry Assignment: Protein Structure Analysis of Human Thymidylate Synthase and Ligand Binding Using the Protein Data Bank

Description

As a graduate student, Professor Angela W. Peters focused on the biochemistry of human thymidylate synthase (TS). Understanding the structure of TS allows students to understand how the ternary complex formation and the structure of the active site is utilized in the treatment of diseases such as cancer (*1*, *2*). In addition, understanding protein structure and protein-ligand binding interactions is also helpful as you introduce students to drug design and the importance of structure and binding. The activity tells the thymidylate story and how enzymes and substrates bind to form a ternary complex. This complex acts as an inhibitor of TS activity. Thus, several classes of drugs have been utilized clinically or experimentally to inhibit TS.

Learning Goals

In this activity, students will:

- Collect and analyze a protein structure data from the Protein Data Bank.
- Visualize the structural features consistent with hydrogen bonding and ligand binding.
- Visualize 3-dimensional images of human TS, specifically the active site, a key target for therapeutics used in the treatment of many types of cancers.
- Understand the importance of protein-ligand binding and how this contributes to developing therapeutic targets.

Equipment Needs

Internet connection and computer access to the Protein Data Bank. The Protein Data Bank is a collection of sequences from several sources. Protein sequences are the fundamental determinants of biological structure and function.

Evaluation

Students can be evaluated by submitting responses to the worksheets for this lesson.

Background

Thymidylate synthase (TS) catalyzes the reaction in which deoxyuridylate (dUMP) is reductively methylated by 5,10-methylenetetrahydrofolate (MTHF) to form thymidylate (TMP) and dihydrofolate (Figure A1). The reaction is the only de novo source of TTP for DNA synthesis and repair and TTP is the only natural nucleotide incorporated into DNA that does not require ribonucleotide reductase (RNR) for its biosynthesis. Because this enzymatic reaction provides the sole intracellular de novo source of dTMP, TS is a critical therapeutic target in cancer chemotherapy (2).

Several classes of drugs have been utilized clinically or experimentally to inhibit TS. The most widely utilized agent is 5-fluorouracil (FUra). FUra is anabolized to a nucleotide analog of dUMP, 5-fluoro-2′-deoxyuridylate (FdUMP), which is a tight-binding inhibitor of TS in the presence of the folate co-substrate, MTHF. A stable inhibitory ternary complex is formed in which the pyrimidine base of FdUMP is bonded covalently to a cysteine residue of TS and to MTHF.

Lesson 1

Images from the Protein Data Bank (PDB) appear in journal articles and in textbooks as 3-dimensional structures. These images are eye-catching and can really help students understand and visualize the structure and function of proteins. This lesson should be completed as a group, following discussions with the Instructor. Prior to the lesson, the Instructor should discuss the role of TS as a therapeutic target for cancer. This is based on the structure of hTS and the ternary complex formation. Students should also learn about amino acids, draw their structures and the functional groups of the amino acids. The students will use the PDB to view the crystal structure of hTS and understand the active site. The Protein Data Bank has been utilized in other engaging chemistry lessons published on the VIPEr (Virtual Inorganic Pedagogical Electronic Resource) website, which is a free resource for chemistry faculty and educators (*11*).

Chemistry professors should have their students complete the following tasks:

- Log into the PDB using the following website:
    - https://www.rcsb.org/
- Use the search bar and type in "crystal structure of human thymidylate synthase".
- Use the mouse to rotate the image and evaluate the crystal structure.
- Review reference (2), which discusses the active site and the important amino acids located at the active site. It also discusses the role of cysteine at the active site, and the ternary complex formation.
- Review Figure A1, which provides the mechanism for TS formation.
- Complete Worksheet 1.

Student Handout

Answer the following questions based on discussions with your group and evaluation of the crystal structure of human thymidylate synthase using the Protein Data Bank. You will need to access the Protein Data Bank at the following website:

Protein Data Bank: https://www.rcsb.org/

1. How many cysteine residues are located in hTS?
2. What is the functional group of cysteine? Draw the structure.
3. What is the definition of a hydrogen bond?
    a. Why is this important in proteins?
    b. What are covalent bonds?
    c. What type of molecules will display hydrogen bonding?

4. Complete Table A1 by clicking on the amino acids located above the 3-D image in the hTS PDB. For each amino acid listed in the table, determine which amino acid forms hydrogen bonds.
5. Select a disease or a symptom and name five common medicines that are used to treat it. What are the side effects?
6. What is a ligand and how does it bind to certain proteins?
7. What are substrates and why are they important in protein binding interactions?
8. What is the importance of the ternary complex formation between hTS, FdUMP and MTHF?
9. Based on the structure of hTS, why is it targeted as a therapeutic agent?

Table A1. Amino Acid Side Chain Characteristics in hTS

Amino acid	*Hydrogen bonds to amino acid*
Ex. Ser234	Asp 231
Cys157	
Cys20	
Gln15	
Thr53	
Asn160	
Asp150	

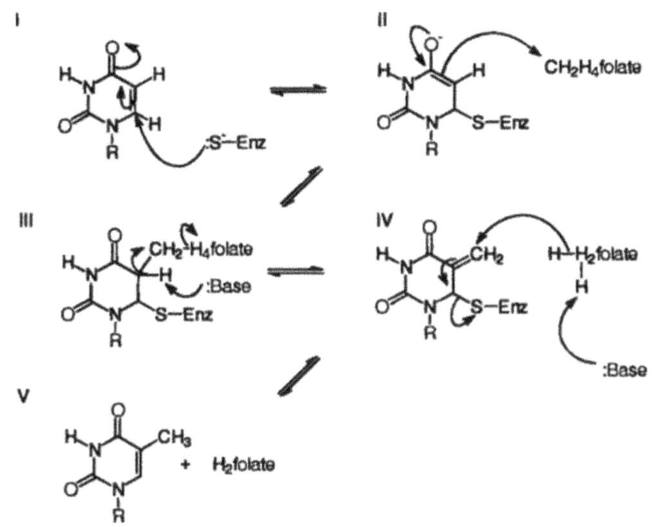

Figure A1. Key features of the reaction mechanism of TS. Reproduced with permission from reference (2). Copyright 1998 American Chemical Society.

References

1. Hoganson, D. K.; Williams, A. W; Berger, S. H. Isolation and characterization of thymidylate synthase deficient human tumor cell lines. *Biochemical Pharmacology* **1999**, *58*, 1529–1537.
2. Williams, A. W.; Dunlap, R. B.; Berger, S. H. A hydroxyl group at residue 216 is essential for catalysis by human thymidylate synthase. *Biochemistry* **1998**, *37*, 7096–7102.
3. Peters, A. W. Teaching Biochemistry at a Minority Serving Institution: An Evaluation of The Role of Collaborative Learning As A Tool For Science Mastery. *Journal of Chemical Education* **2005**, *82*, 571–574.
4. Peters, A. W.; Mani, D.; Rasathurai, S.; Greene, M. The Effectiveness of Supplemental Instruction and Technology in Increasing Student Performance in Mathematics. *Journal of Women and Minorities in Science and Engineering* **2006**, *12* (1), 1–9.
5. Peters, A. W.; Tisdale, V. A. Paradigm Shift: Infusing Research into Teaching and Learning at Claflin University. In *Infusing Undergraduate Research into Historically Black Colleges and Universities Curricula* (Diversity in Higher Education, Volume 17); McClinton, J.; Melton, M. A.; Jackson, C. R.; Engerman, K., Eds.; Emerald Group Publishing Limited, 2015, pp.47–61.
6. Peters, A. W.; Tisdale, V. A.; Swinton, D. High-Impact Educational Practices that Promote Student Achievement in STEM. In *Broadening Participation in STEM (Diversity in Higher Education*, Volume 22); Wilson-Kennedy, Z. S.; Byrd, G. S.; Kennedy, E.; Frierson, H. T., Eds.; Emerald Publishing Limited, 2019, pp. 183–196. ISBN: 9781787569089
7. Pew Research Center. *Diversity in the STEM workforce varies widely across jobs*. 2018. https://www.pewsocialtrends.org/2018/01/09/diversity-in-the-stem-workforce-varies-widely-across-jobs/, (Accessed Nov 10, 2020).
8. National Science Board (NSB). *Science and Engineering Indicators 2018*. NSB 2018-1. National Science Foundation: Arlington, VA. 2018. https://www.nsf.gov/statistics/2018/nsb20181/ (Accessed Feb 12, 2021).
9. *World Economic Forum. The Global Gender Gap Report 2017*. Geneva Switzerland: World Economic Forum. https://www.weforum.org/reports/the-global-gender-gap-report-2017 (Accessed Feb 12, 2021).
10. Nestor-Baker, N.; Kerka, S. *Recruitment and Retention of Underrepresented Students in STEM Fields*. The Ohio State University: Columbus, OH, 2009. http://mediaimpact.issuelab.org/resources/10301/10301.doc (Accessed Feb 12, 2021).
11. Jamieson, B. *Protein Data Bank: A Beginner's Guide*. IONiC VIPEr, https://www.ionicviper.org/web-resources/protein-data-bank-beginners-guide (Accessed Feb 27, 2021).

Chapter 10

Mentoring in Life and Science

Renã A. S. Robinson*

Department of Chemistry, Vanderbilt University, Nashville, Tennessee 37235, United States

***Email: rena.as.robinson@vanderbilt.edu.**

This chapter focuses on the impact of mentoring on the career trajectory of analytical chemist, Dr. Renã AS Robinson. She earned her doctoral degree in analytical chemistry from Indiana University in 2007 and pursued a career in academia. She started as an Assistant Professor at the University of Pittsburgh and currently is a professor at Vanderbilt University. Through her life experiences in education and science and her research emphasis on the use of analytical chemistry to understanding aging, Alzheimer's disease, and health disparities she has realized the true impact of mentoring.

Introduction

Mentoring is a term that I did not become familiar with until maybe high school, or even until undergraduate years, but it is a benefit that I have had the opportunity to experience throughout my entire life. Excellent mentoring is absolutely critical for students pursuing careers in science, technology, engineering, and mathematics (STEM), especially when those students are from underrepresented backgrounds. As an African American and Black female chemist, I have learned first-hand and through experiences of mentees and colleagues of the challenges that come with a lack of quality mentorship in STEM. My own STEM journey, has had such a profound impact, that I founded a company Purposeful PhD, LLC that seeks to transform the graduate student journey. Effective mentoring is a necessary component for all students, and especially for underrepresented students in STEM, to successfully navigate their journey.

Family and Life Mentoring

Before I dive into mentoring along my educational journey in STEM it is important that I reflect on the concept of mentoring within my own family. My first mentors were in my household as I grew up. My mother, a dietician at Norton's hospital and my father a mechanic and former employee of L&N Railroad Company, together, set a high bar of expectations for my three brothers and I at an early age and offered whatever they could to help us navigate our educational journey. Some of these expectations were spoken directly to us. For instance, I heard stories from aunts and uncles that my father was exceptional in math growing up and that math must be somewhere in my genes.

Other expectations were unspoken. My mother was an example of what hard work looked like as she became the primary breadwinner in our family when my father's health took a turn and he passed away from heart complications resulting from diabetes and kidney disease when I was in middle school. The spoken expectations in our house were that we (my siblings and I) should do well in school and do something meaningful with our lives. Even outside of my home, right next door and in the neighborhood, there were other adults who also placed high expectations on me once they heard that I was doing well in school. By affirming for me that I was smart and praising me for doing well in school, I became very accustomed to this form of mentoring and today am truly grateful and appreciate that support. It's a simple fact that there is no one person that can mentor in this life and help us to achieve all that we are destined for, but, in fact, it is a collection of mentors that are critical for our journey.

There is formal mentoring, that we tend to think of mostly when it comes to our careers, but what I call life mentoring which helps instill the principles that you need to be successful in anything, is equally as important. My brothers (Figure 1) mentored me on being independent and knowing how to take care of myself and of others, whether that was through cooking a family meal or walking to the store to get groceries for the household. They showed me what it meant to have responsibility and how to manage responsibility with grace. This started when my brother Terry had children and I became an Auntie at six years old and began my first babysitting jobs when I was around ten years old. My brother Jeffrey taught me the power of knowledge and relying on faith, which is the foundation from which I operate. My youngest brother, Maurice, taught me that being the smartest or the best, is not always most important, because there always are others who deserve to have the same accolades and opportunities as well. They taught me that mentors have a responsibility to share their wisdom and place value on helping others.

Figure 1. Photo from 2009. On Renã's wedding day to Antonio W. Robinson (not shown), shown here are her three brothers, bottom left (Maurice Sowell), top left (Terral Sowell, Sr.), top right (Reverend Jeffrey Sowell). Photo courtesy of Renã A. S. Robinson.

Outside of my home and neighborhood, I spent most of my young life at church, school, or in afterschool programs, where I was exposed to many mentors. Honestly, there are too many to name and I will not name them for fear of missing someone. At church-Brown Memorial Christian Methodist Episcopal in Louisville, KY- my pastors, Dr. Charles King and Reverend Raymond Williams, trusted me to pray, speak, or read as a part of worship services and to take on leadership roles with the youth. Many women in the church poured their hearts into me, teaching me how to be a young lady, reinforcing the high bar of expectations set by my mom regarding education and doing well in and out of school. They took the time to be present and care for me like I was their own daughter. The lessons I learned in church around faith, leadership, sharing, giving, organization, grace, humility, compassion, responsibility, respect, and friendship from my pastors, women in the church, leaders, and my peers, helped shape me in ways that one mentor alone could not have done. These were all life mentors that showed me the importance of mentors pouring into mentees without having any expectation on receiving anything in return. Mentors serve others because it's the right thing to do.

K-12 Education in School and in Afterschool Programs

Early mentoring in childhood or the lack of mentoring, helps to shape the trajectory for young people in STEM. The effective mentoring I received during my formative K-12 years played a significant role on where I am today in my career. My neighborhood was predominately Black and thus, most of the students in the elementary school near us, Whitney M. Young Elementary, were also Black. I recall two teachers that were critical to me in these years as mentors: Ms. Ford and Ms. Hayden. In the year I was in Ms. Ford's class, she always held me to a high standard and was willing to correct me when needed. Perhaps she did this because she was the teacher and teachers do this, but I wonder if it was because she saw something in me and did not want me to fall into an unproductive path.

Mrs. Hayden, another teacher at Whitney M. Young, mentored me in a way that was compassionate, soft, and nurturing. She took an important interest in me that went beyond the classroom and cared about me as an individual. She often talked in depth with my mom when she would see her in the neighborhood grocery store, and still took interest in me and my growth as a person and life-long learner well after I left Whitney M. Young Elementary. These mentors taught me the importance of stepping in to help mentees correct their path when you see them getting off track, and that mentors can serve a life-long purpose.

My science memories go back to middle school when I studied in Mr. White's classroom. He made a strong impression because he was a Black male science teacher and was the first person to show me that science could be fun. He used his classroom to create funny and fun experiences for his students. He ate a big bowl of Frosted Flakes cereal every morning that he would finish before getting the lesson started. He took such hold of his classroom that he grabbed my attention with his lessons and science demonstrations. I loved being in his class and at that time it did not strike me as odd that he was a Black male science teacher.

In high school at Louisville Central High Magnet Career Academy, I encountered another Black science teacher, Mrs. Braxton, who taught biology and genuinely wanted students to learn science and make connections with it in our daily lives. Mr. Emory, my chemistry teacher, made science fun in every single class and while I did not see myself as a scientist at this time in my life, I recognized that I enjoyed the hands-on nature of the class and mastering what most considered difficult concepts.

Outside of school, my mom ensured I was involved in many afterschool programs from my elementary to high school years. The Whitney M. YOUNG Scholars (*1*) and the Math and Science for Minority Students (MS^2) (2) Programs absolutely were two of the most critical and influential programs and experiences that I had during my K-12 education (3). The Scholars Program started when I was in 7th grade and ended with my high school graduation. This was a college readiness program that required that I spend days after school and weekends taking specialized classes with peers from other high schools across Louisville, KY and summers going to college campuses across the state for two-week residential experiences. Beyond the leadership and student teachers who simply ushered us through the program, I had the honor to experience invested mentorship firsthand from Mr. Benjamin Johnson, Mr. Emory Lee, Mr. Andre Guess, Mrs. Wilma Claybourne, and Dr. Samuel Robinson (President of the Lincoln Foundation and founder of the Whitney M. YOUNG Scholars Program). These individuals were not just interested in my success in the program but, were genuinely dedicated to and interested in ensuring that I, and all of my peers were setup to succeed in every aspect of our lives. Collectively, they taught me that mentors have a heart for and care deeply for their mentees with a long-term view in mind.

I was blessed to receive similar mentorship at MS^2 by program leaders and summer instructors. However, in this program, the most profound mentorship I received came from the ~50 other students that were in the program with me. MS^2 was held during three high school summers at Phillips Academy Andover, and was designed to promote and accelerate math and science education for African American, Hispanic/Latinx and Native American students from all over the United States. This was a 5-week summer residential experience that placed students in advanced and intensive math and science courses. Many of them put us ahead when we returned to our high schools the following fall. I absolutely loved this program and the breath of fresh air and support that came from seeing other students of color excel in math and science. My peers were brilliant (every single one) and while smarts distinguished this group, the most important thing I learned from them was to appreciate people and respect everyone's culture. My peers, through free and fun encounters, sometimes heated, but jovial debates about who had the correct answer to a discrete math problem, or working together to design a cultural showcase, taught me that mentors can be peers who are where you are or are slightly ahead of you, but have learned lessons that you have not yet mastered.

Indiana University Doctoral Research

My career trajectory, in retrospect, was destined to be in math or science and academics, based on my constant connection to learning through my formal education, afterschool, and summer programs. Early on, I had dreams of becoming a cardiac surgeon to help prevent others from dying from heart failure, which was ultimately what took the life of my father in middle school. Later, I realized the commitment that would come with that path and leaned toward the pursuit of an entrepreneurial career in cosmetics. This realization came my junior year at the University of Louisville, when I shifted to a major in Chemistry with a concentration in Business. While, I had this goal in mind, I did not have a clear plan or know the best route to achieve it. While I had mentors in college who helped me navigate my course schedule from semester to semester, I was missing a mentor in chemistry who could help me clarify what my next steps should be. In fact, I'll share that I did not really understand what graduate school was until the semester after I graduated (in 3 ½ years) from the University of Louisville. From this experience, I learned that a lack of mentorship can slow down or prevent certain opportunities from being taken advantage of because you just don't know what you don't know.

I attended graduate school at Indiana University (IU) with the initial intent of working as hard as I could to start a cosmetics company at the end of my doctoral degree. However cosmetics were not an option in the program and there was no clear path to doing this. Dr. David Clemmer, my graduate advisor and life-long mentor, convinced me that I could still get to where I was going, but I could do it through a project in his laboratory that used analytical chemistry to measure proteins to understand aging. My graduate work focused on the development of ion mobility and mass spectrometry (MS) approaches to study changes in *Drosophila melanogaster* (fruit flies). I was interested in the entire complement of proteins present in cells, tissues, organs, or the entire body of the fruit fly which puts the work in the field of proteomics (*4*). Proteomics was still in its infancy while I was a graduate student so I learned a lot about how to interpret mass spectra of peptides and proteins. Proteins are molecules necessary for cellular function and activity. They are composed of chains of amino acids (we have 20 naturally occurring amino acids) and can vary in length, sequence, and 3-dimensional structure (*5*).

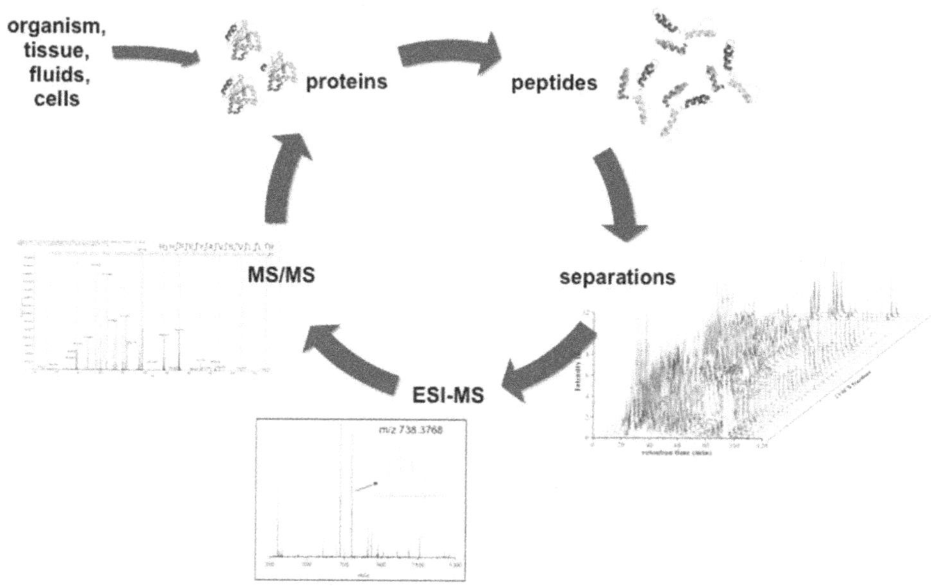

Figure 2. Example workflow of bottom-up proteomics.

In bottom-up proteomics, proteins are extracted from tissues (in my case I was using fly heads) and digested using chemical enzymes such as trypsin which cleave C-terminal at arginine and lysine residues (*6, 7*) (Figure 2). These peptide mixtures are easier to ionize than intact proteins, using electrospray ionization (ESI), an ionization technique that garnered the Nobel Prize in Chemistry in 2002 (*8*). Upon ESI, peptides in small droplets can receive multiple charges and upon desolvation and coloumbic explosion, become highly charged (generally protonated) gas-phase ions before entering the mass spectrometer. I was fortunate to work with several mass spectrometers while I was a graduate student which included, a linear ion trap MS, a custom in-house ion mobility-time-of-flight mass spectrometer, and a high-resolution Fourier Transform ion cyclotron resonance mass spectrometer. The mass spectrometer detects gas-phase peptide ions as electrical current that can be converted and calibrated to determine the mass-to-charge (m/z) ratio of the peptide ion. Through tandem MS (MS/MS), peptide ions can be collisionally activated in the mass spectrometer, and broken apart into fragments. These fragments are also detected and used to determine the amino acid sequence of the peptides. Database search algorithms are used to determine the identification of proteins present in the sample mixture. This proteomics approach was used many times in my dissertation work and applied to study the biology of aging in fruit flies (9), which although an invertebrate species, shares a significant genetic homology with humans.

I believe the most profound result from my dissertation work was using proteomics to determine that a protein, prophenoloxidase, was declining with age in the fruit fly (*10, 11*) (Figure 3). This is a protein that is critical for defense response and wound healing. I measured hundreds of proteins in fruit fly heads and many of them did not change with increasing age in heterozygous wild-type flies, such as the protein paramyosin (Figure 3). However, the protein, prophenoloxidase, had a dramatic decline with age in the flies and this prompted us to ask the question "Is prophenoloxidase necessary for the flies to survive?" In order to answer this question, we compared the survivorship of two heterozygous wild-type fly strains with a mutant strain that did not have prophenoloxidase activity. The results (Figure 3) clearly show that without prophenoloxidase being active the flies do not live as long. I was amazed at this result that allowed me to better understand the connection of my proteomics analytical measurements with biology of aging in a living organism. This was a moment in my journey that was celebrated.

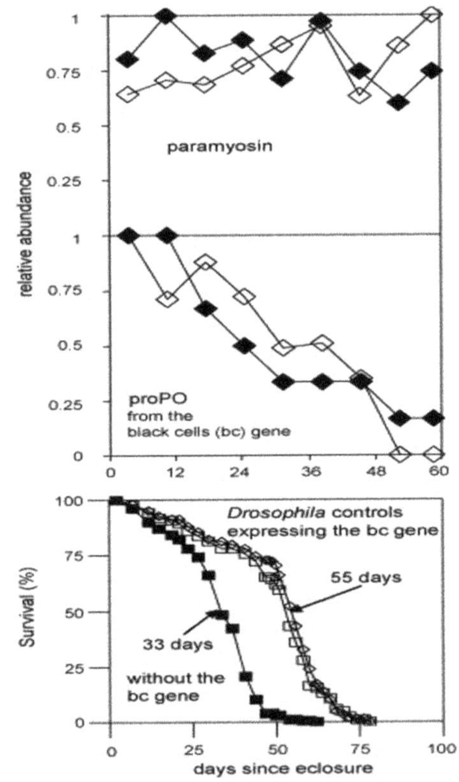

Figure 3. Liquid chromatography-tandem mass spectrometry (LC-MS/MS) and LC-ion mobility-MS data of proteins from Drosophila melanogaster heads. The data traces in the top show the relative protein expression of paramyosin and prophenoloxidase proteins. The bottom plot shows the survival curve for two heterozygous populations of Drosophila melanogaster and a homozygous population that is a mutant which lacks activity of prophenoloxidase enzyme. Reproduced with permission from Reference (11). Copyright 2021 Elsevier.

Dr. Clemmer helped me to dream as a graduate student, and encouraged me to think about how this protein could potentially make it into an anti-aging cosmetics product. As a mentor, Dr. Clemmer took the time to teach me about entrepreneurship and through many conversations about writing, research, and excellence, taught me the lesson that mentors can inspire their mentees to think and dream beyond their career goals and plans.

By the end of my graduate career I truly believed I had earned my rights as a certified mass spectrometrist and a beginning proteomics expert. There is much to add about this six-year journey especially as it was not an easy one and I was surrounded by peers and professors at IU that did not look like me. In fact, it wasn't until I graduated that the then Black president of IU, Adam W. Herbert, shared with me while crossing the stage (as my family was shouting loudly and cheering from the bleachers) how proud I should be of my historic accomplishment of being a Black female PhD in Chemistry from IU. Of course, I was keen on being the only one throughout graduate school as was often the case in many environments I had been in previously throughout my educational journey. However, in graduate school it became more of a burden and a weight at the time because aside from my graduate advisor it was not as straightforward to find mentors in my department that I could talk to that would understand this aspect of my training and unique experiences. A couple of years into my program, Dr. Michael Edwards, a professor of public and environmental affairs and chemistry, (and a Black professor), took the time to check on me and how I was really doing in graduate school. He took it a step further sometimes for myself and other Black students, by also being the voice that spoke up for us when needed so that our issues or complaints could be heard. From this experience, I began to learn that mentors have the power to use their voices to be advocates for their mentees.

Independent RASR Laboratory Research

I left graduate school unsure about what my path would be but very clear that it would not be in academia. I also knew that I was not driven just by money and I did not want to be stuck in an industrial career that would pay well but make me miserable on the job. In retrospect, had I been exposed to other mentors in academic or industrial careers, I may have gained a different and positive perspective on choosing either of those career paths. I was fortunate to be mentored as a postdoctoral fellow by Dr. D. Allan Butterfield in the Department of Chemistry at the University of Kentucky and to receive a Lyman T. Johnson and a UNCF/Merck postdoctoral fellowship to join his laboratory. It was in his laboratory that I expanded my applications of proteomics and MS to study the oxidation of proteins in aging and in particular, Alzheimer's disease. There, I used mammalian models such as mice, rats and eventually expanded into the analysis of human tissues (*12–14*), which resonated with my innate passion to help people with my skills in analytical chemistry. In Dr. Butterfield's laboratory, I witnessed the interaction of clinical researchers with basic science researchers. This helped connect for me the possibility that my research could directly impact people. Dr. Butterfield's approach to mentoring was very transparent and genuine and he was especially bold about his support of women and people of color in science. He had received a Presidential Award for Excellence in Science, Mathematics, and Engineering Mentoring prior to my joining his laboratory. He showed me that mentors can come from backgrounds different from their mentees and only need to have the heart to help.

It was during my postdoc when I realized that part of my calling or purpose in life was to use my chemistry skills and knowledge to help others who were underserved and historically underrepresented in STEM to navigate the graduate school journey, and make it into academic and professional roles in science.

I began to reflect on my entire academic experience and recognize the lack of representation of women and Black scientists in my educational journey. Additionally, I recognized that the disparities in education impacted not only one's career journey but also their overall health. I took a deep dive in the literature to learn about health disparities in Alzheimer's disease amongst African American/

Black and Latinx adults in the US (*15*, *16*). I decided to pursue a career in academia to provide opportunities for historically underrepresented students in science and to use my analytical chemistry research to develop a program that would ultimately provide a significant benefit to society.

While at the University of Pittsburgh, I built a research program that developed advanced quantitative proteomics techniques to multiplex many samples in a single MS analysis. These techniques, called combined precursor isotopic labeling and isobaric tagging (cPILOT) (*17–21*), were applied to study aging and Alzheimer's disease. Through this technology development and application, I have had the pleasure to mentor many undergraduate, graduate, and a few high school students, postdoctoral associates, and scientists. As much as we have worked to advance our science and have it make an impact in the field of analytical chemistry, I have worked to be and continue to become a better mentor. I have realized that each student is unique and has their own set of strengths and areas for growth and improvement. Communication with each mentee is key and can vary depending on how mentees respond to communication styles. However, I ensure that as much as possible I am accessible to my mentees. I always encourage mentees to think about the big picture of their training regardless of their career stage (undergraduates, graduates, postdoctoral, or junior faculty) and use tools such as semester strategic plans and standing one-on-one meetings to help mentees stay on track and adjust priorities as needed. Listening to mentees is important in order to understand what drives and motivates them and also to learn what are their personal and professional aspirations. It is from this insight that I try to best support them and connect them to individuals in my network or resources that can further support them on their own individual journey.

Most importantly, I recognize that none of my mentees are me. As a mentor, I've recognized that it's not my job to create another version of myself or try to mold someone else to do or be what I have become, but rather to use my knowledge to help my mentees become the best versions of themselves and give guidance on possible routes to get to where they are going.

My research at Vanderbilt University continues to develop advanced proteomics and MS techniques to propel our understanding of health disparities in Alzheimer's disease, especially amongst African American and Black adults (*22*). Our research is highly interdisciplinary and it is our goal to accelerate our understanding of disease and disparities so that we can reduce the burden and disproportionate number of sufferers from Alzheimer's disease. While I am trained as an analytical chemist, my willingness to reach outside of my discipline has helped me to understand the vastness of the research problems we seek to solve. It has also shown me that is important for me to continue to be mentored by other colleagues who can help me navigate this aspect of my journey for my own career. Mentoring happens throughout your entire lifetime.

NOBCChE

Reflecting back on my journey in higher education, it is critical to highlight a resource that transformed my thoughts about scaling mentorship. As the only Black graduate student in my department for a couple of years at IU, I recognized that I needed to seek mentorship and gain access to role models outside of my department and University. Dr. Clemmer supported my first attendance of the National Organization for the Professional Advancement of Black Chemists and Chemical Engineers (NOBCChE) Annual Conference. I remember attending this conference and getting exactly what I had been missing in both undergraduate and graduate school: the opportunity to see a massive number of Black scientists at all levels in a single location. My life as a graduate student was transformed, because I knew that I was not the only one anymore, as there were many others that looked like me all across the country, in fact throughout the world.

Through my involvement with NOBCChE, I gained countless mentors who took time to talk with me, call and email me, and walk me through many things, some as simple as building up a stellar resume, to presenting my science better, to determining the next steps of my research efforts after scientific presentations. Of all the NOBCChE mentors, I could name, I'll just share two that were critical to my pursuing a career as a professor: Dr. Isiah Warner, from Louisiana State University, an analytical chemist and distinguished professor, and Dr. Aleeta Powe, from the University of Louisville, also an analytical chemist and professor. They were exactly in the position I was considering and because of NOBCChE, I had direct access to reach out to them and ask for real guidance and honesty about choosing this career path. Through their mentorship over the years, I have learned that mentors have to be transparent and real about their experiences in order to truly help mentees, be prepared for what may come their way later.

Something special also happened as I continued to be involved with NOBCChE. Through attending annual conferences, and participating in chapters at different institutions, I have had the opportunity to increase the scale of my own mentoring of students and early career professionals. As a NOBCChE member, Board-member at large (2016-2019), and President-Elect (2019-2021), I have been able to meet and talk with hundreds of talented, gifted, and extraordinary students and colleagues through the years. When I can listen, make a connection, or can share a nugget or two that I've picked up along the way, I am more than willing to do so. At this stage in my career and life, I recognize that it is my privilege and duty to be a small part of what helps someone else achieve success as a scientist and in life.

Chapter Summary

Mentoring has been a natural part of my journey both in receiving and giving mentorship. However, since my early years as a faculty member, I have focused on the development of ideas and strategies to figure out how some of this can be done at scale, and how systems can be setup to provide better mentoring to all students. Mentoring is a huge part of what will ensure historically underrepresented students in STEM gain access to the knowledge and resources that they need to be equally represented in science. I founded and started Purposeful PhD, LLC, in 2018 with a mission to ensure the success and well-being of graduate students by transforming their journey (23). I have learned through working with my clients that mentors cannot take for granted how many individuals they can help with just a little bit of time and encouragement.

Finally, this chapter also includes a literature review assignment (Appendix) based on one of my laboratory's publications in the ACS journal, *Analytical Chemistry*. Chemistry professors can implement this assignment in their analytical chemistry or more advanced chemistry courses for student engagement.

Acknowledgments

I would like to thank every single person whether a mentor formally or informally, mentee, and individual that has poured into my life for any reason, season, or lifetime. Each of you has helped make the difference in my journey and for that I am grateful.

Appendix: Sample Teaching Assignment

Bottom-Up Proteomics and Mass Spectrometry

Description

This lesson plan highlights the chemistry contributions of Dr. Renã AS Robinson, an Associate Professor of Chemistry at Vanderbilt University. In this activity, students will have the opportunity to observe a quantitative proteomics experiment and learn how to calculate peptide masses and observe mass shifts from labeling in high-resolution mass spectra.

Learning Goals

There are various learning goals for this assignment:

1) Students will determine the MW of a peptide using experimental *m/z* values.
2) Students will understand how chemical additions to peptides impact *m/z* values.
3) Students will apply analytical concepts related to mass spectrometry, isotopic labeling, isobaric tagging, and bottom-up proteomics in a peer-reviewed journal article.

Equipment Needs

Internet connection, calculator, and computer access.

Implementation Notes

The instructor will need to provide the students access to watch a short video to complete this assignment. The video will help students to visualize and see some of the steps explained in the journal article and answer the questions in the student handout. The video can be accessed at the following url: https://www.jove.com/v/55406/enhanced-sample-multiplexing-tissues-using-combined-precursor.

After watching the video, students should read the following two peer-reviewed articles:

- Evans AR, Gu L, Guerrero R, Jr., Robinson RAS*. Global cPILOT Analysis of the APP/PS-1 Mouse Liver Proteome. Proteomics: Clin Appl. **2015**; 9:872-84.
- King C, Dudenhoffer J, Gu L, Evans AR, Robinson RAS*. Enhanced Sample Multiplexing of Tissues Using Combined Precursor Isotopic Labeling and Isobaric Tagging (cPILOT). J. Vis. Exp. **2017**; 123:e55406.

Evaluation

Students will be evaluated on their responses from the following questions on the Student Handout.

Student Handout: Literature Discussions and Video on Proteomics

Literature Review and Video

The video and journal articles below address a quantitative proteomics technique developed to increase sample multiplexing in proteomics analyses of tissues from mammalian models of Alzheimer's disease. The video can be accessed via the following url: https://www.jove.com/v/55406/enhanced-sample-multiplexing-tissues-using-combined-precursor.

After watching the video, read the following two peer-reviewed articles:

- Evans AR, Gu L, Guerrero R, Jr., Robinson RAS*. Global cPILOT Analysis of the APP/PS-1 Mouse Liver Proteome. Proteomics: Clin Appl. **2015**; 9:872-84.
- King C, Dudenhoffer J, Gu L, Evans AR, Robinson RAS*. Enhanced Sample Multiplexing of Tissues Using Combined Precursor Isotopic Labeling and Isobaric Tagging (cPILOT). J. Vis. Exp. **2017**; 123:e55406.

After reading the two journal articles and watching the video, answer the following questions for a class discussion:

1. What are the building blocks of a protein?
2. Why are enzymes, such as trypsin, used in proteomics experiments?
3. Given the *m/z* ratio of a peak in a mass spectrum obtained using electrospray ionization, the MW of the peptide can be obtained using the following formula:

   $$\text{MW (peptide)} = m/z * z - z^{*}\text{m}_\text{p}$$

   whereby, z is the charge state of the peak, and m_p is the mass of a proton (1.007825 Da). Using this information calculate the MW of the triply-charged peptide shown in the mass spectrum in Figure A1C at *m/z* 759.068.
4. Since the peptide peak in the spectrum (Figure A1C) is dimethylated in two positions, what is the mass of the light labeled peptide without any dimethylation?
5. How are tandem mass tags able to label six samples such that they can be combined and analyzed simultaneously in the Orbitrap mass spectrometer?
6. Now using information learned about the TMT reagents and tags in the article, calculate the MW of the peptide peak in Figure A1A from the peak shown at *m/z* 831.442.
7. How does the MW of the peak change when you remove the mass that corresponds to a dimethyl group and a TMT group?

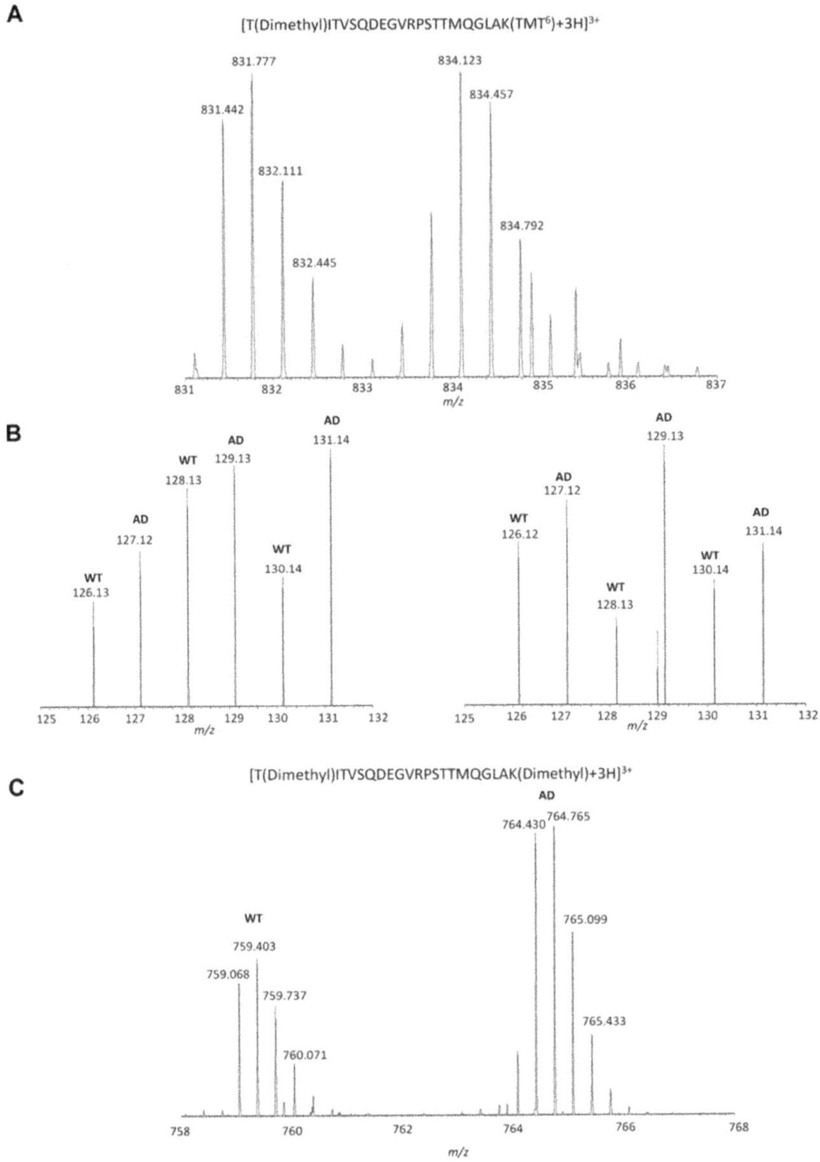

Figure A1. (A) Example mass spectrum for the peptide [T(dimethyl)ITVSQDEGVRPSTTMQGLAK-(TMT6)+3H]3+ which belongs to 3-ketoacyl-CoA thiolase B (Uniprot Acc. No. Q8VCH0) with (B) left, HCD-MS3 reporter ion spectra from light (m/z = 831.442) and right, heavy (m/z = 834.123) peptide peaks. Only the low m/z region is shown. (C) MS spectrum for the peptide [T(dimethyl)ITVSQDEGVR-PSTTMQGLAK(dimethyl)+3H]3+ in the precursor dimethylation experiment. The light (m/z = 759.068) and heavy (m/z = 764.430) peptide peaks correspond to WT and AD samples, respectively. Reproduced with permission from Reference (19). Copyright 2021 John Wiley and Sons.

References

1. *Lincoln Foundation*. https://lincolnfdn.org/ (Accessed March 14, 2021).
2. *Andover Math and Science for Minority Scholars (MS^2) Program*. https://www.andover.edu/about/outreach/ms2 (Accessed March 14, 2021).
3. Robinson, R. A. S. A Divinely Ordered Path in *Mom the Chemistry Professor: Personal Accounts and Advice from Chemistry Professors who are Mothers*, 2nd Edition, Springer Publishing Company, Switzerland AG, Woznack, K., Charlebois, A., Cole, R., Marzabadi, C.,Webster, G., Eds., DOI: 10.1007/978-3-319-78972-9 2018.
4. Gygi, S. P.; Aebersold, R. Mass spectrometry and proteomics. *Curr Opin Chem Biol.* **2000**Oct, *4* (5), 489–94doi: PMID: 1100653410.1016/s1367-5931(00)00121-6.
5. *Wikipedia "Protein" Page*. https://en.wikipedia.org/wiki/Protein (Accessed March 14, 2021).
6. Taraszka, J. A.; Kurulugama, R.; Sowell, R. A.; Valentine, S. J.; Koeniger, S. L.; Arnold, R. J.; Miller, D. F.; Kaufman, T. C.; Clemmer, D. E. Mapping the Proteome of Drosophila melanogaster: Analysis of Embryos and Adult Heads by LC-IMS-MS Methods. *J Proteome Res.* **2005**, *4*, 1223–37.
7. Taraszka, J. A.; Gao, X.; Valentine, S. J.; Sowell, R. A.; Koeniger, S. L.; Miller, D. F.; Kaufman, T. C.; Clemmer, D. E. Proteome Profiling for Assessing Diversity: Analysis of Individual Heads of Drosophila melanogaster Using LC-Ion Mobility-MS. *J Proteome Res.* **2005**, *4*, 1238–47.
8. Press release. NobelPrize.org. *Nobel Media AB 2020.* Thu. 31 Dec 2020. https://www.nobelprize.org/prizes/chemistry/2002/press-release/ (Accessed March 14, 2021).
9. Sowell, R. A. *Proteome measurements for developing hypotheses about the biology of Drosophila*. ProQuest Dissertations Publishing 2007 Indiana University, Bloomington, Indiana
10. Sowell, R. A.; Hersberger, K. E.; Kaufman, T. C.; Clemmer, D. E. Examining the Proteome of Drosophila Across Organism Lifespan. *J Proteome Res.* **2007**, *6*, 3637–47.
11. Valentine, S. J.; Kurulugama, R. T.; Bohrer, B. C.; Merenbloom, S. I.; Sowell, R. A.; Mechref, Y.; Clemmer, D. E. Developing IMS-IMS-MS for rapid characterization of abundant proteins in human plasma. *Int J Mass Spectrom.* **2009**, *283*, 149–60.
12. Robinson, R. A S.; Lange, M. B.; Sultana, R.; Galvan, V.; Fombonne, J.; Gorostiza, O.; Zhang, J.; Warrier, G.; Cai, J.; Pierce, W. M.; Bredesen, D. E.; Butterfield, D. A. Differential expression and redox proteomics analyses of an Alzheimer disease transgenic mouse model: effects of the amyloid-β peptide of amyloid precursor protein. *Neuroscience (Amsterdam, Neth).* **2011**, *177*, 207–22.
13. Robinson, R. A. S.; Joshi, G.; Huang, Q.; Sultana, R.; Baker, A. S.; Cai, J.; Pierce, W.; St. Clair, D. K.; Markesbery, W. R.; Butterfield, D. A. Proteomic analysis of brain proteins in APP/PS-1 human double mutant knock-in mice with increasing amyloid β-peptide deposition: Insights into the effects of in vivo treatment with N-acetylcysteine as a potential therapeutic intervention in mild cognitive impairment and Alzheimer's disease. *Proteomics.* **2011**, *11*, 4243–56.
14. Aluise, C. D.; Robinson, R. A. S.; Cai, J.; Pierce, W. M.; Markesbery, W. R.; Butterfield, D. A. Redox Proteomics Analysis of Brains from Subjects with Amnestic Mild Cognitive Impairment Compared to Brains from Subjects with Preclinical Alzheimer's Disease: Insights into Memory Loss in MCI. *J Alzheimer's Dis.* **2011**, 23, 257–69.
15. Alzheimer's Association. 2017 Alzheimer's disease facts and figures. *Alzheimers Dementia* **2017**, *13*, 325–373.

16. Chin, A. L.; Negash, S.; Hamilton, R. *Diversity and disparity in dementia: the impact of ethnoracial differences in Alzheimer disease*. Alzheimer Dis. Assoc. Disord., Lippincott Williams & Wilkins, Philadelphia, Pennsylvania 25 (2011), pp. 187-195, 10.1097/WAD.0b013e318211c6c9.
17. Robinson, R. A. S.; Evans, A. R. Enhanced Sample Multiplexing for Nitrotyrosine-Modified Proteins Using Combined Precursor Isotopic Labeling and Isobaric Tagging. *Anal Chem (Washington, DC, U S)*. **2012**, *84*, 4677–86.
18. Evans, A. R.; Robinson, R. A. S. Global combined precursor isotopic labeling and isobaric tagging (cPILOT) approach with selective MS3 acquisition. *Proteomics.* **2013**, *13*, 3267–72.
19. Evans, A. R.; Gu, L.; Guerrero, R., Jr.; Robinson, R. A. S. Global cPILOT analysis of the APP/PS-1 mouse liver proteome. *Proteomics: Clin Appl.* **2015**, *9*, 872–84.
20. Gu, L.; Robinson, R. A. S. A simple isotopic labeling method to study cysteine oxidation in Alzheimer's disease: oxidized cysteine-selective dimethylation (OxcysDML). *Anal Bioanal Chem.* **2016**, *408*, 2993–3004.
21. King, C.; Dudenhoffer, J.; Gu, L.; Evans, A. R.; Robinson, R. A. S. Enhanced Sample Multiplexing of Tissues Using Combined Precursor Isotopic Labeling and Isobaric Tagging (cPILOT). *J. Vis. Exp.* **2017**, *123*, e55406.
22. Stepler, K. E.; Mahoney, E. R.; Kofler, J.; Hohman, T. J.; Lopez, O. L.; Robinson, R. A. S. Inclusion of African American/Black adults in a pilot brain proteomics study of Alzheimer's disease. *Neurobiology of Disease* **2020**, *146*, 105129. ISSN 0969-9961, https://doi.org/10.1016/j.nbd.2020.105129
23. *Purposeful Ph.D., LLC*. www.purposefulphd.com/ (Accessed March 14, 2021).

Chapter 11

In the Rhythm of Agency

Leyte Winfield*

Department of Chemistry and Biochemistry, Division of Natural Sciences and Mathematics, Spelman College, Atlanta, Georgia 30314, United States

***Email: lwinfield@spelman.edu.**

This chapter provides a personal account of my intellectual achievements and passion for nurturing the success of black women in chemistry. With the exception of residencies, fellowships, and sabbaticals, my career was birthed and sustained in the southern and southeastern region of the United States. I was educated in New Orleans, Louisiana, and began my career as a post-doc in Tallahassee, Florida, before moving to Atlanta, Georgia, where I have built a 17-year career in academia.

The Woman in Black

Carrie May Weems is a photographer noted for her work, which examines the historical and social complexities of inequality, access, and agency. In the *Museum Series*, Carrie Mae Weems examines the juxtaposition of herself, the artist and the subject, in the foreground of prominent institutions such as the Galleria Nazionale D'Arte Moderna in Rome and the Louvre Museum in Paris (*1*). Adorned in a matronly black dress, Weems positions herself at a distance where it appears, she ponders entering the institution or has another reason to pause before accessing the museum. I have experienced the pause of the woman in black looking in on a world that seems distant from my norm. As I consider Anna Julia Cooper's notion of "when and where I enter" (*2*), I regard this pause captured by Weems as an opportunity to process how one gains access to an environment and the lives impacted by one's presence in that environment. With each new endeavor along my career journey, I encountered this moment of reflection, a moment to celebrate my accomplishments and the legacy of agency held by black women in STEM.

It Started at the Bend in the River

The capital of Louisiana, Baton Rouge, is situated at the bend in the Mississippi River, 80 miles northeast of New Orleans. Legend has it that the capital city received its name from a stick that was painted red and found on the riverbanks near its current location. Itti Houma, the Choctaw word for red pole and known by locals as istrouma, was translated by the French as Baton Rouge. The city is a combination of Creole and Cajun, Spanish and French giving rise to a somewhat eclectic African American culture. Within this culture, I developed a love for science, beginning with mathematics, flirting with computer science, then finding my heart in chemistry. My mom taught me multiplication facts when I was four years old (*3, 4*). While I look back and wonder what she was thinking, I recognized that her instilling in me an understanding of basic arithmetic gave me the foundation needed to excel in STEM (science, technology, engineering, and mathematics). My pathway towards STEM was affirmed by individuals who declared, "she is going to be a doctor," or "she is so good at math." My adolescent mind believed the pronouncements and naturally gravitated towards activities that fed my interest in STEM. My budding curiosity for experimentation was evident in my need to improve cosmetics and personal care products, most notably mixing lipstick colors or combining moisturizes to improve the lubricating effect. The curiosity led me to chemistry, and I entered high school, fully committed to the idea of becoming a chemist. I followed that passion earning a B.S. in chemistry from Dillard University and a Ph.D. in organic chemistry, with a research concentration in drug design and synthesis, from the University of New Orleans. The academic foundation I obtained at these institutions not only built my intellectual capacity in the chemical enterprise, but also planted in me a desire to contribute to science in service to the greater good.

University of New Orleans Doctoral Research in Drug Design

One of the first things you do in a Ph.D. program is to interview potential research advisors or rotate through several labs to determine the best fit for your interest and temperament. Selecting projects that could impact the lives of African Americans is a core value threaded through all of my professional endeavors. I realized early in my career that I wanted to do research that would be easy to explain to a lay audience. Therefore, being accepted into Dr. Mark Trudell's research group was somewhat serendipitous to these core values. My project focused on designing novel compounds that could govern the action of neurotransmitters and responded to current events, e.g., the infamous war on drugs, of relevance to my community. Through the project, I explored the relationship between a compound's structure and its ability to reduce an individual's desire for cocaine, thereby alleviating the addiction.

GBR 12909 was the lead molecule for the study due to its slow onset of activity, low abuse potential relative to that of cocaine, and ability to inhibit dopamine uptake at the presynaptic neuron, Figure 1. We aimed to create a fast-acting, long-lived compound having utility as a pharmacological probe to study the effects of drug transport and entry into the brain (*5*). We believed that the rate of entry into the brain and the timeframe of activity had some bearing on the likelihood that the user would become dependent on the compound. Lipophilic compounds can bind to proteins and be absorbed into other fatty tissues in the body. Compounds with large lipophilicity values can have a slow rate of entry into the brain. Therefore, two series of GBR 12909 derivatives were designed. The structural features of each coincide with the compounds having lower lipophilicities than the parent compound GBR 12909. The derivatives, 2,6-dioxopiperazine and 1,3,5-diazine analogs, are shown in Figure 1.

A total of 20 compounds were synthesized (*6, 7*). While the compounds were less lipophilic than GBR 12909, the analogs were less efficacious than the lead molecule towards inhibiting the uptake of dopamine. In addition to the synthesis and biological assays, I developed rudimentary models to rationalize the design of the compounds, as shown in Figure 2. The computational analysis of medicinal compounds and their biological interactions was an emerging field at the time. The multitude of computational tools has expanded since then, allowing me to incorporate these techniques into my current research and curricular efforts.

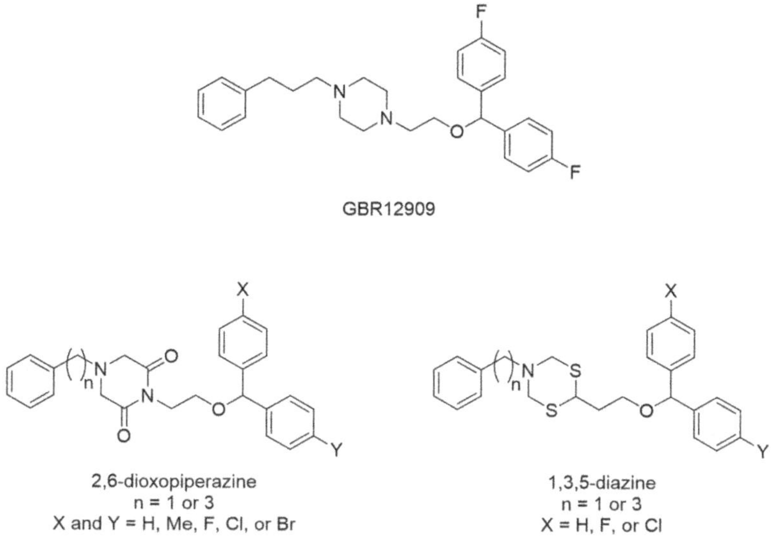

Figure 1. Structure of the lead compound and the series of derivatives from my dissertation.

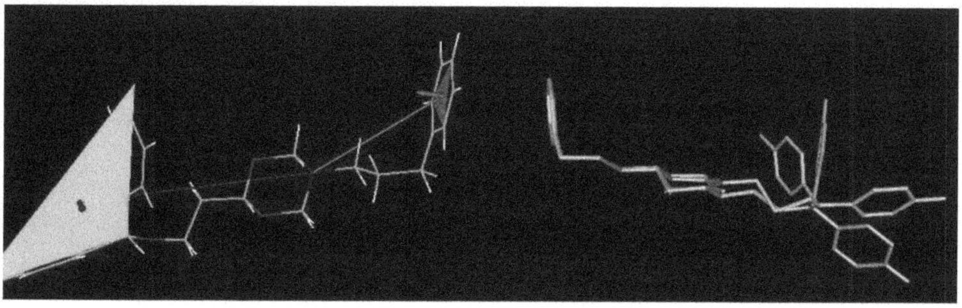

Figure 2. Structural overlay of the dioxopiperazine and diazine analogs.

In graduate school is where I received my first opportunity to mentor. After my research advisor agreed to the idea, I recruited an undergraduate from Dillard University. I was both excited and intimidated. I was excited about the prospect of "reaching-back" to select a student from my alma mater and providing the individual with an experience that could enrich their development. I was intimidated by the fact that the quality of the student's experience was my responsibility. I was not confident that I had created a project that would keep the individual's attention; neither was I sure that I could guide the person's success on the project. Fortune would have it that the individual selected was Christy Reid, a talented student whose ingenuity enabled her to work with little direction. Her outcomes became a critical piece of my project and involved developing a procedure to calculate the compounds' lipophilicity. She also completed the assays and related calculations. It was a memorable first mentoring experience. She even introduced me to Destiny's Child and Alicia Keys. Without judgment, she simply remarked, "You should check them out. I think you will like them." I regret losing contact with Christy. The experience, however, piqued my interest in pursuing a career in academia.

In addition to research, my time in graduate school was marked by becoming a mom in the first year of the program. Finding balance is critical to professional and personal success. However, it took time for me to realize this to be true and balance my life appropriately. As a result, I did not enjoy the level of socialization that most did in their twenties. I had to choose focus over fun, a choice that was necessary but unfulfilling in retrospect. My primary forms of social engagement included being involved in my church community and doing STEM outreach whenever the opportunity presented itself. Adding to this, I was a commissioned officer in the U.S. Army Reserve. The training brought a welcome distraction from the lab, and I enjoyed serving my country. Nevertheless, the demands of being a military officer, full-time graduate student, and single mom left little time for other activities. Managing these three aspects of my life developed in me a level of discipline that continues to inform my career success.

Establishing Independent Lab Research as an Academic

After completing my Ph.D., I became a post-doc at Florida A&M University in the College of Pharmacy. I left my post-doctoral position after only nine months to accept the assistant professorship at Spelman College. The position was made for me, allowing me to leverage my passion and talent for chemistry to train women of African descent who will change the world. The research I developed at Spelman College centered around reproductive cancers that disproportionately impact the African American community. The project involved designing novel tricyclic molecules (di-substituted benzimidazoles, otherwise known as the LLW molecules) and evaluating their therapeutic potential against prostate, breast, and ovarian cancers. To achieve these aims, computational methods were used to create visual models that depict how the benzimidazole-based molecules interact with biological targets. The features were further optimized to enhance their therapeutic potential.

The synthesis and purification of these molecules involve a four-step process, illustrated in Figure 3. Step 1 is performed under microwave conditions to produce a series of diaryl nitro amine as orange to deep-red, prismatic solids. Step 2 was conducted under microwave conditions to produce near-quantitative yields of the product with high purity. The resulting diamine is obtained as a purple wax-like solid. The purification of the molecule formed in Step 4 is tedious but produces the desired substance as a white powder in modest yields.

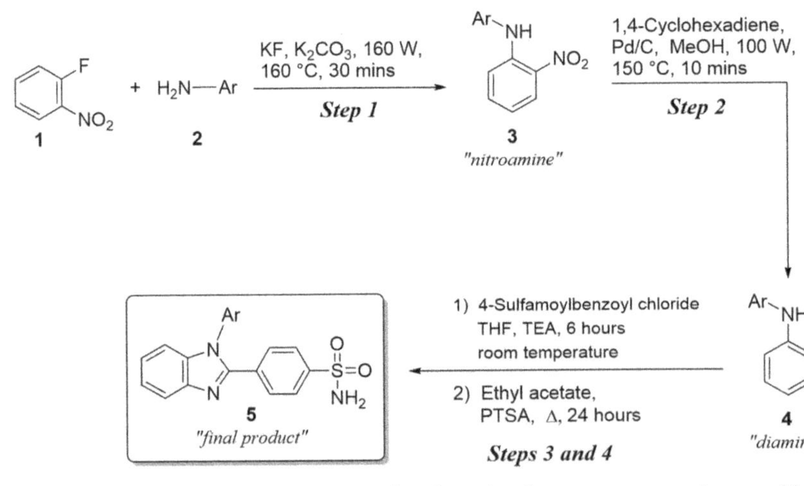

Figure 3. Synthesis of aryl-substituted benzimidazole molecules. Ar = para-substituted benzene, benzyl, phenoxy, naphthalenyl, and anthracenyl.

The benzimidazole molecules have demonstrated the ability to inhibit the growth, survival, and migration of breast carcinoma (MCF-7 and MDA-MB-231), prostate carcinoma (PC-3 and LNCAP), and ovarian cystadenocarcinoma (SKOV-3) cell lines. The molecules synergistically inhibited the growth of human glioma brain cancer cells LN18 when given in concert with a nonlethal dose of sulfasalazine (*8–11*). The latter finding contributes to research on overcoming drug resistance, a common issue with current therapeutic options. The molecules' activity is attributed, in part, to its ability to inhibit COX-2 and estrogen activity, among other biological targets. This type of polypharmacology is emerging as a sought-after feature in pharmaceutics (*12*, *13*). The LLW molecules have shown promise in being repurposed to treat other diseases, thereby expanding their utility. The molecules' biological profile garnered interest, leading to six patents covering eleven unique tricyclic structural frameworks for research on cancer and inflammation-related diseases, Figure 4 (*14–19*). The molecular library offers hope for a potentially safer, orally available alternative to current therapeutic options that could be given independently or in combination with other drugs.

Figure 4. Tricyclic framework for patented molecules with structural variations occurring at the Ar and A–E points.

Research as a Training Ground for Cultivating STEM Identity

The solutions sought throughout our world require input from all constituents, including those from groups underrepresented in STEM. Research experiences led by scholars who can serve as role models are critical for propelling women of color's longevity in STEM-related careers. In *Seeing the Invisible Scientist*, Dr. Carmine Leggett, a former research student in my laboratory, says:

> *I saw you [Dr. Winfield] being a leader in the sciences. That image was very empowering to see. I used that memory as fuel in graduate school, because honestly, continuing on, I didn't see many Black female researchers or had any Black female research mentors. In graduate school, I was one of the few African Americans. Seeing my struggles with some coursework, a professor strongly encouraged me to leave the sciences. I ignored his advice and went on to do great work. I even graduated on the Dean's list. I continued in the program not to prove him wrong, but because I love science, and I was determined to earn my doctorate. Most importantly, I knew I was capable of doing the work* (*20*).

Reflecting on her time as my research advisee, Chyree Batton notes in an interview for *Chemical & Engineering News* that "Winfield really anchored me there [at Spelman College] and helped focus my energy so that when I did graduate, I had a purpose, I knew what I was doing, and I could give back" (*21*). What is striking about the quotes is the assertion of their science identity, sense of belonging, and belief that they have something valuable to add to their disciplines. I take pride in knowing that these students credit my mentorship, in part, for affirming these identities.

In the laboratory, I train students to navigate techniques and instrumentation and nurture their sense of inquiry through the research process. Students explore reaction conditions (e.g., solvents, temperature, and time), microwave and conventional heating, and purification methods (e.g., workups, recrystallization, and chromatography). They also create computer-generated models of biological interactions (i.e., kinase and DNA binding), calculate physicochemical characteristics and ADMET (absorption, distribution, metabolism, excretion, and toxicity) properties, and developed mathematical models and graphical analysis of computed properties. Their research accomplishments have been nationally recognized. For instance, students' work has been twice named Outstanding Undergraduate Research by the Council of Undergraduate Research. The

Council selects less than 100 undergraduates from disciplines across academia and from institutions throughout the United States. Honorees are invited to Capitol Hill to present their work to members of Congress. Students' research efforts have also won awards at competitions hosted by the National Science Foundation and the American Association for the Advancement of Science at their annual Emerging Researchers National Conference. Several individuals from my lab have placed first at the Spelman College Research Day. Of the thirty undergraduates I have mentored, seven have earned doctorates in dentistry, medicine, or pharmacy. Others have pursued their post-baccalaureate education at institutions such as The Rockefeller University, Albert Einstein College of Medicine, University of Connecticut, and Louisiana State University, with five having received STEM doctorates, four having earned STEM master's degree, and four currently completing doctoral studies. The spirit of inquiry and the analytical skills afforded through undergraduate research experiences are transferable to other fields. Leveraging this, one of my past mentees earned an MBA and is excelling in an executive marketing career. Another earned a Juris Doctorate and is pursuing a career in intellectual property law.

Bringing the My Research to the Curriculum

The early translation of my research into a curricular module focused on developing an inquiry-based project for the Organic Chemistry II Laboratory (22). The first step of the synthesis outline in Figure 3 was modified, allowing students to explore substituent reactivity in a nucleophilic aromatic substitution, Figure 5. The class was split into three groups, each receiving a different 2-halonitrobenzene. The groups shared reaction data and used the information to determine the leaving group trend for the reaction.

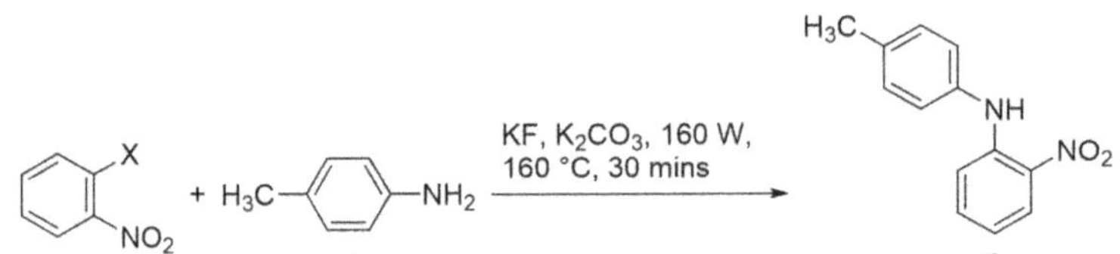

1 2 3

Figure 5. Reaction scheme for the nucleophilic aromatic substitution reaction used in the inquiry-based laboratory. X = F, Cl, or Br.

I have used the computational aspects of my research to create modules for undergraduate organic and medicinal chemistry lecture courses. In organic chemistry, an iSpartan enabled activity was implemented to introduce the basic concepts of alkane conformation. Students manipulated three-dimensional structures within the app and calculate conformational energies and dihedral angles (23). The information was used to answer questions related to substituent interactions. Computational modeling was infused into the medicinal chemistry course through a series of activities featuring conformational analysis, structural overlays, binding studies, and pharmacophore analysis. The activities were compiled in a self-published lab manual. In one activity, students examined the binding interactions of indinavir that governs the drug's ability to inhibit the HIV protease (24). Likewise, medicinal compounds are routinely used to discuss the properties of organic structures, particularly functional groups, in the organic chemistry lecture course. In the second-semester organic chemistry course, this approach is useful for addressing the significance of reactions. I infused the reactions used in my research laboratory into the conversation (i.e., the reduction reaction shown in Figure 3, step 2). Using this information also illustrates for students that information from the textbook has implications on work being done at Spelman College.

Transitioning to Broadening Participation and Discipline-Based Education Research

Broadening Participation

The need for valuing "diversity work" was the focus of the article "Systemic Racism in Higher Education, an open letter on the climate in STEM for faculty of color" (*25*). The article aligns with my motivation for conducting scholarship that empowers women of color in STEM-related careers. The solutions sought throughout our world require input from all constituents, including those from groups underrepresented in STEM. There have been numerous endeavors to address the retention of these individuals in STEM disciplines. My work in the area adds to these efforts by characterizing initiatives at primarily undergraduate and minority-serving institutions such as Spelman that ensure these individuals' longevity in the STEM workforce. Spelman College, in particular, has experienced exceptional success in fostering bachelor's degree attainment in all areas, including STEM. My work contributes to Spelman's models of success by highlighting the narratives of academically mobile women of color. The work uncovers how focal individuals describe (a) their agency and other markers of their success and (b) the structures/entities that have supported their success.

My engagement in this field ensures a first-person perspective on Spelman's unique learning community as well as that of other minority-serving institutions. The first-person perspective allows us to move from being the subjects of the investigation to the curators of the knowledge. I am creating symposia and developmental opportunities that have made an impact beyond STEM and academia, as demonstrated by my residencies at the Rockefeller Foundation's Bellagio Center and the Women's International Study Center. Both opportunities were relevant to my endeavors around promoting gender equity in STEM while translating the importance of such work in realizing more inclusive and sustainable economies. My work in this area has been captured in several publications (*26–30*) and resulted in my receipt of a visiting professorship at the University of Michigan.

Discipline-Based Education Research

As higher education embraces online and web-enhanced pedagogies, it is necessary to characterize the impacts of these strategies on learning behaviors among women of color at small liberal arts environments. Mainly, it is of interest to understand how students engage with learning resources and respond to collaborative learning activities. Therefore, I am characterizing agency in interactive and peer-learning spaces, such as flipped classrooms. Through these activities, I promote gender equity in science careers by developing culturally relevant initiatives and curricula that support the productive engagement of minorities and women in academia. In this way, my discipline-based education research complements my broadening participation work, and at times, the two are indistinguishable.

Utilizing design-based research methodologies, I created a flipped learning environment informed by the community of inquiry (CoI) framework, Figure 6 (*31*). Hallmarks of CoI include an overlap among online learning, face-to-face instruction, and social interaction that promotes peer-peer and peer–instructor discourse. Complementing the CoI framework, agency and associated affective traits (e.g., effort, mindset, motivation, and efficacy) are considered drivers of self-regulated learning. These traits are crucial to a student's ability to engage effectively in a flipped or hybrid learning environment.

To date, we have characterized student agency and how such agency correlates to student engagement in the course (*32–36*). The fact that most students demonstrate high career and grade motivation in these studies is not a novel observation. The cohort of students in these studies also exhibits high self-determination with slightly lower intrinsic motivation and self-efficacy. We examined these components of motivation and have identified a correlation between the frequency with which students employ higher-impact study strategies and students' level of self-determination and grade motivation. We have also found that students report having prepared to engage in in-class activities using the instructor's online resources. Students have indicated that actively participating in course activities enhanced their understanding of chemical concepts. In further evaluating interactions during collaborative and peer-led learning activities, we are capturing the social and cognitive presence of the CoI in a way that uniquely reflects the Spelman learning experience.

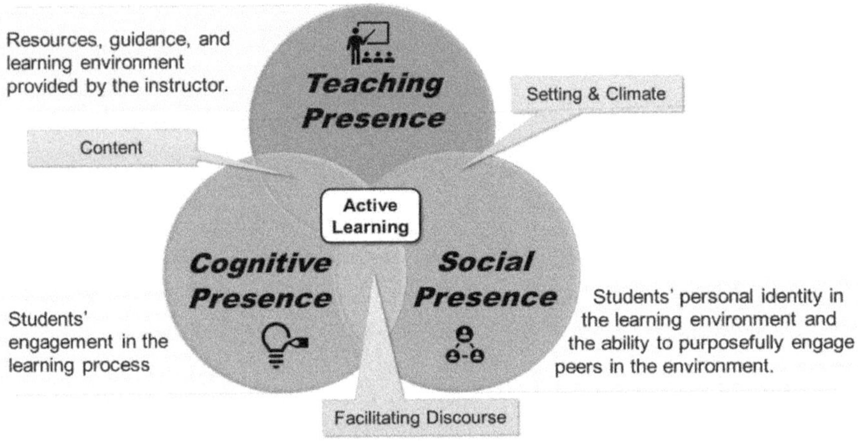

Figure 6. The CoI framework.

Chapter Summary

As I conclude this chapter, the nation is grappling with a global pandemic, calls to address racial inequities, and an extreme political divide. Described in the preceding text are scholarly activities that have converged to empower Black women's equitable participation in STEM careers. The activities have included engaging students in authentic research and adapting this work to demonstrate for students the relevance of information presented in the curriculum. I am also characterizing elements of student science identity that are informed by one's agency. These efforts, in turn, ensure that our discipline contains the type of gendered knowledge needed to affect women's lives and accurately depict the lived experiences of Black women in STEM.

Appendix: Sample In-Class Activity

Intermolecular Forces and Physical Properties

Description

This activity was developed for the first-semester organic chemistry course. In this activity, students work with electrostatic potential maps, ball and stick structures, and tube structures. The electrostatic potential maps are examined to explain the intermolecular forces experienced by each compound. The information is used to describe relative melting and boiling points and solubilities. The activity also contains real-life examples relevant to intermolecular forces.

Cultural Relevance

In 2018, the Black hair market grossed more than $2.51 billion, according to Essence magazine (37). Nielsen reports that consumer spending in the African American community proportionally exceeds their representation in the U.S. population (38). For instance, of the $3.04 billion spent on personal soap and bath needs, the Black community purchased $1.3 billion of these products. Further, Black consumers spent an estimated $54.4 million on ethnic hair care products compared to the $63.5 million total spent in this area. Complimenting the consumer impact of Blacks in the personal care market, YouTube content developers contributed an estimated 45,000 channels to beauty in 2019. This module examines the intermolecular forces that underlie the properties and formulations of personal care and medicinal products.

Learning Goals

Students will be able to:

1. Explain what is represented by the colors of an electrostatic potential map.
2. Identify dipoles and the degree of polarity within a molecule.
3. Associate specific intermolecular forces with a given structural component (i.e., functional groups).
4. Describe the trend in physical properties for a series of compounds.

Resources Needed

No additional resources are needed. However, students may build and manipulate the activity in iSpartan or similar structure viewing software.

Implementation Notes

Students should receive a lecture describing intermolecular forces and their relevance to physical properties. The lectures can include culturally relevant examples of the importance of intermolecular forces. In my course, we address the surface area by describing the desire to assume a fetal position to warm oneself on a cold night. We also consider the energy required to straighten natural hair, considering if a given head of hair is long or short. This analogy again points to the surface area using an example relevant to the Black community.

Students can recreate molecules using an app such as iSpartan to manipulate the structures in three-dimension. If three-dimensional models such as ball and stick and tube structures have not been introduced in the course, students should receive instruction on interpreting these structures. If students have access to iSpartan, students can be asked to build the surface map and compare what they see for each structure.

The activity can be completed in one class period or given as a homework assignment. I typically allow students to complete the activity in groups.

Evaluation

Students are evaluated based on their responses to the worksheets for this lesson. Different binding modes can be used for the final question to create different cases for cultural relevance.

Draw a depiction of the strongest intermolecular force experienced by propanol ($CH_3CH_2CH_2OH$, B.P. = 97 °C) and methyl ethyl ether ($CH_3OCH_2CH_3$, B.P. = 7.4 °C). Considering the two structures and your drawing, why is the boiling point of alcohols higher than that of ethers having the same number of carbons.

Explain the differences and the similarities between the structures shown in Figure A1. Include what you see and what the illustration represents in your explanation. Based on your assessment of the differences, indicate which molecule has the highest melting point.

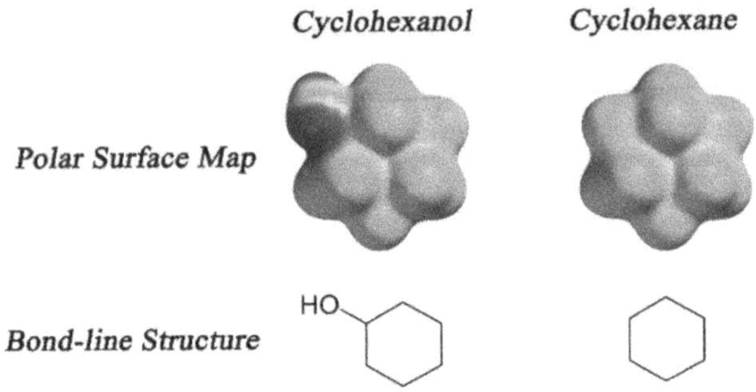

Figure A1. The surface maps and bond-line structures for cyclohexanol and cyclohexane.

There is a large market for beauty products that reduce the signs of aging by reducing the appearance of wrinkles or acne. A key ingredient in many of these products is vitamin A. One might say that vitamin A is to fighting wrinkles as vitamin C is to fighting colds describes literature, frameworks, and methodologies appropriate for broadening participation and DEIR work? However, vitamin C can be found in a number of orally consumable products (for instance, orange juice, a water-based solution), and vitamin A is found in creams and serums (which are primarily hydrocarbons). Using Figure A2 which shows the electrostatic potential map of each molecule, explain why.

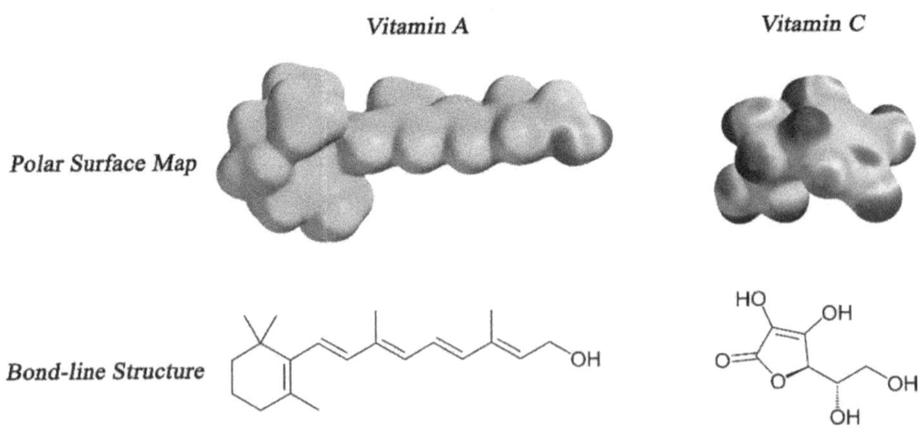

Figure A2. The surface maps and bond-line structures for Vitamin A and Vitamin C.

Celebrex is an FDA approved drug for the treatment of pain. The molecule acts by blocking the activity of an enzyme called Cyclooxygenase 2 (COX-2), an enzyme responsible for the swelling, redness, and pain response associated with an injury. The drug is routinely prescribed for the chronic pain associated with Lupus and Sarcoidosis, which disproportionately impacts the African American community. Drugs interact with various biomolecules via IMF. Figure A3 depicts interactions between Celebrex (ball and stick structure) and COX-2 (tube structure).

Interaction A is a hydrogen bond. Is the atom of the Celebrex molecule acting as a hydrogen bond acceptor or donor in the interactions? What type of IMF is represented by interaction B?

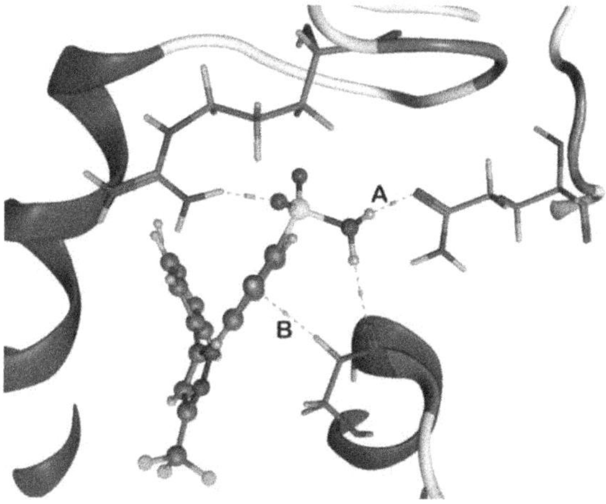

Figure A3. Interactions between Celebrex and the COX-2 protein. The dashed lines represent IMF between two groups or atoms. Atoms are represented by ball or the tips of the tubes. Atom colors are as follows: dark grey–carbon, light grey–hydrogen, red–oxygen, yellow–sulfur, blue–nitrogen, green–fluorine.

Acknowledgments

This work has been funded in part by the National Science Foundation HBCU-UP Targeted Infusion Project Award No. HRD-1332575 and Division of Undergraduate Education Award No. 1626002.

References

1. Weems, C. M. *Museums 2006*. http://carriemaeweems.net/galleries/museums.html (accessed Feb 5, 2021).
2. Lemert, C. Anna Julia Cooper: The Colored Woman's Office. In *The Voice of Anna Julia Cooper: Including a Voice from the South and other important essays, papers, and letters*, Lemert, C.; Bhan, E., Eds. Rowman & Littlefield Publishers, Inc.: Lanham, MD, 1998.
3. Brown, J. E. *African American Women Chemists in the Modern Era*. Oxford University Press: New York, 2018.
4. Winfield, L. Finding Rhythm. In *Mom the Chemistry Professor : Personal Accounts and Advice from Chemistry Professors who are Mothers*, Woznack, K.; Charlebois, A.; Cole, R.; Marzabadi, C.; Webster, G., Eds.; Springer International Publishing: Cham, 2018; pp 539-551. DOI: 10.1007/978-3-319-78972-9_40.
5. Winfield, L. *Synthesis, Structure, Dopamine Transporter Affinity and Octanol/Water Partition Coefficients of Novel, Less Lipophilic GBR 12909 Analogs*. Dissertation, University of New Orleans, New Orleans, LA, 2002.

6. Winfield, L.; Izenwasser, S. E.; Wade, D.; Trudell, M. L. Synthesis and Dopamine Transporter Binding Affinity of 2, 6-Dioxopiperazine Analogues of GBR 12909. *Med. Chem. Res.* **2002,** *11* (2), 102–115.
7. Winfield, L.; Zhang, C.; Reid, C.; Stevens, E.; Trudell, M.; Izenwasser, S.; Wade, D. Synthesis, lipophilicity and structure of 2,5-disubstituted 1,3,5-dithiazine derivatives. *J. Heterocycl. Chem.* **2003,** *40*, 827–832.
8. Payton-Stewart, F.; Tilghman, S. L.; Williams, L. G.; Winfield, L. Benzimidazoles Diminish ERE Transcriptional Activity and Cell Growth in Breast Cancer Cells. *Biochem. Biophys. Res. Commun.* **2014,** *450* (4), 1358–62. DOI: 10.1016/j.bbrc.2014.06.130.
9. Yerokun, T.; Winfield, L. L. LLW-3-6 and Celecoxib Impacts Growth in Prostate Cancer Cells and Subcellular Localization of COX-2. *Antican. Res.* **2014,** *34* (9), 4755–4759.
10. Gregory-Bass, R.; Winfield, L. Preliminary Analysis of Anti-proliferative, Apoptotic, and Anti-migratory Effects LLW-3-6 in SKOV-3 Ovarian Cystadenocarcinoma Cell Line. *Lett. Drug Des. Discov.* **2020,** *17*, 1319–1324. DOI: 10.2174/1570180817666200129142949.
11. Yerokun, T.; Winfield, L. L. Celecoxib and LLW-3-6 Reduce Survival of Human Glioma Cells Independently and Synergistically with Sulfasalazine. *Antican. Res.* **2015,** 35 (12), 6419–6424.
12. Bolognesi, M. L. Harnessing Polypharmacology with Medicinal Chemistry. *ACS Med. Chem. Lett.* **2019,** *10* (3), 273–275. DOI: 10.1021/acsmedchemlett.9b00039.
13. Winfield, L. L.; Payton-Stewart, F. Celecoxib and Bcl-2: Emerging Possibilities for Anticancer Drug Design. *Future Med. Chem.* **2012,** *4* (3), 361–383. DOI: 10.4155/fmc.11.177.
14. Winfield, L. *Syntheses and Anti-Proliferative Effects of Benzimidazole Derivatives*. U.S. Patent 7,947,723 B2, May 24, 2011.
15. Winfield, L. *Syntheses and Anti-Proliferative Effects of 1-Aryl-substituted Benzimidazole Derivatives*. U.S. Patent 8,106,084, January 31, 2012.
16. Winfield, L. *Synthesis and Anti-Proliferative Effect of Substituted Imidazo[4,5-C]Pyridine Compounds*. U.S. Patent 8,329,728, December 11, 2012.
17. Winfield, L. *Synthesis and Anti-Proliferative Effect of Substituted Imidazo[4,5-B]Pyridine Compounds*. U.S. Patent 8,334,302, December 18, 2012.
18. Winfield, L. *Synthesis and Anti-Proliferative Effect of Substituted Purine Compounds*. U.S. Patent 8,357,691, January 22, 2013.
19. Winfield, L. *Synthesis and Anti-Proliferative Effect of Substituted Imidazo[4,5-B]Pyrazine Compounds*. U.S. Patent 8,362,019, January 29, 2013.
20. Winfield, L.; Sally-Nicholson, C., Seeing the Invisible Scientist. In *Alumni Profiles and Perspectives on STEM and the Impact of Undergraduate Research at Spelman College*. Spelman College: Atlanta, GA, 2018; p. 45. https://radar.auctr.edu/islandora/object/sc.fac.pubs%3A2018_winfield_leyte (accessed Feb 5, 2021).
21. McDaniel, J. ACS Scholar Alumna: Chyree Batton. *Chemical & Engineering News* **2020**.
22. Winfield, L. Nucleophilic Aromatic Substitution, A Guided Inquiry Laboratory Experiment. *Chem. Educator* **2010,** *15*, 110–112.
23. Winfield, L.; McCormack, K.; Shaw, T. Using iSpartan To Support a Student-Centered Activity on Alkane Conformations. *J. Chem. Educ.* **2019**. DOI: 10.1021/acs.jchemed.8b00145.

24. Winfield, L.; Gregory-Bass, R.; Campbell, J.; Watkins, A. Characterizing Ligand Interactions in Wild-type and Mutated HIV-1 Proteases. *J. Comp. Sci. Educ.* **2014**, 5 (1), 1–9. DOI: 10.22369/issn.2153-4136/5/1/1.
25. Barber, P. H.; Hayes, T. B.; Johnson, T. L.; Márquez-Magaña, L. Systemic Racism in Higher Education. *Science* **2020**, *369* (6510), 1440. DOI: 10.1126/science.abd7140.
26. *Growing Diverse STEM Communities: Methodology, Impact, and Evidence*; Winfield, L. L.; Thomas, G.; Watkins, L. M.; Wilson-Kennedy, Z. S., Eds.; ACS Symposium Series 1328; American Chemical Society: Washington, DC, 2019.
27. Jackson, K. M.; Winfield, L. L. Realigning the Crooked Room: Spelman Claims a Space for African American Women in STEM. *Peer Review: Emerging Trends and Key Debates in Undergraduate Education* **2014**, *16* (2), 9.
28. Wilson-Kennedy, Z. S.; Payton-Stewart, F.; Winfield, L. L. Toward Intentional Diversity, Equity, and Respect in Chemistry Research and Practice. *J. Chem. Educ.* **2020**, *97* (8), 2041–2044. DOI: 10.1021/acs.jchemed.0c00963.
29. Winfield, L., An Evolving Voice in Leadership. In *Women Called to Lead: Empowering Women of Color in Academic Leadership*, Engerman, K.; Luster-Teasley, S., Eds. Fielding University Press: Santa Barbara, CA, 2017; pp 118—128.
30. Winfield, L. Finding Rhythm. In *Personal Accounts and Advice from Chemistry Professors Who are Mothers*, 2nd ed.; Woznack, K.; Charlebois, A.; Cole, R.; Marzabadi, C.; Webster, G., Eds. Springer: New York, 2018; pp 539–551. DOI: 10.1007/978-3-319-78972-9_40.
31. Garrison, D. R.; Vaughan, N. D., *Blended Learning in Higher Education: Framework, Principles, and Guidelines*; John Wiley & Sons: Hoboken, NJ, 2008.
32. Winfield, L. In *Community-based interactive engagement in an organic chemistry course*, ICERI 2015: 8th International Conference of Education, Research and Innovation, Seville, Spain, Seville, Spain, 2015; pp 2502-2508.
33. Winfield, L.; Hibbard, L. *Lessons from Chemists on Improving Learning Behaviors and Outcomes*. 2015. SoTL Commons Conference. 63. https://digitalcommons.georgiasouthern.edu/sotlcommons/SoTL/2015/63 (accessed Feb 8, 2021).
34. Winfield, L.; Hibbard, L. B.; Jackson, K. M.; Johnson, S. S., Cultivating Agency through the Chemistry and Biochemistry Curriculum at Spelman College. In *Broadening Participation in STEM*, Emerald Publishing Limited: 2019.
35. Winfield, L.; Sanders, S.; Thrill, C. Self-Regulated Learning in Organic Chemistry. *Critical Conversations and the Academy, NETWORK: A Journal of Faculty Development* **2019**.
36. Winfield, L.; Sanders, S.; Thrill, C. Reflecting on Teaching Presence While Transitioning to Remote Instruction. *Science Education and Civic Engagement* **2020** (2), 79–81.
37. Holmes, T. E. Feature: The Industry That Black Women Built. *Essence*, October 22, 2019. https://www.essence.com/news/money-career/business-black-beauty/ (accessed Feb 5, 2021).
38. *Black Impact: Consumer Categories Where African Americans Move Markets 2018*. https://www.nielsen.com/us/en/insights/article/2018/black-impact-consumer-categories-where-african-americans-move-markets/ (accessed Feb 5, 2021).

Chapter 12

Lessons Learned

Sibrina N. Collins*

Lawrence Technological University, The Marburger STEM Center, Southfield, Michigan 48075, United States

***Email: scollins@ltu.edu.**

How are chemists portrayed in the classroom and in society? Who does chemistry, and why? This chapter explores common themes emerging from the stories celebrating the chemists in this book, and discusses some of the steps that can be taken to broaden the image of a chemist.

Who Does Chemistry, and Why?

About four years ago, I read an interesting op-ed published in *The Physics Teacher* focused on addressing equity in the physics classroom (*1*). "Teaching physics should include teaching the culture of physics—the traditions, assumptions, and habits of mind that are part of the discipline," argues Moses Rifkin, a high-school physics teacher in the Pacific Northwest. Rifkin also developed a curriculum with a unit addressing a key fundamental question—who does physics, and why? Rifkin's course allows students to understand the demographics of physicists in the United States (*2*). Reflecting on the important equity work of Rifkin in physics, I came to the realization that in my chemistry classroom, I too was focused on a central question, who does chemistry, and why?

Recently, Schinske et al. (*3*) investigated teaching strategies to improve the science identity of students enrolled in a first-year course taught at a community college with a diverse population. "Science identity intersects with and is influenced by a person's other identities such as race, gender, or religion (*4*)." "From television shows and movies to websites, news articles, and advertisements, the media recurrently conveys images of who does science, more often than not showcasing a relatively narrow view of science and scientists. Setting the media aside, perhaps we need look no further than our own classrooms to understand the ways scientists are portrayed," write the researchers (*3*). A key point that they emphasize is that educators can make a significant impact on broadening the image of a scientist in their classrooms with students. More importantly, the researchers conclude that a lack of diverse images of scientists can negatively impact students from historically underrepresented backgrounds and hinder them from continuing in science. Thus, *African American Chemists: Academia, Industry, and Social Entrepreneurship* is focused on two ideas, namely a 'chemistry identity' and, who does chemistry, and why?

Emerging Chapter Themes—Effective Mentoring, Chemistry Identity, and Technical Credibility

Developing a 'science identity' or specifically a 'chemistry identity' is obviously an important idea or concept, which connects all of the chemists highlighted in this book. Common themes emerging from the stories in this book include effective mentoring, which leads to establishing a 'chemistry identity,' and developing important skills such as critical thinking and creativity. Essentially, a strong 'chemistry identity' and skill sets—which can be nurtured through effective mentoring—are interrelated and enable chemists to establish technical credibility within an organization (Figure 1). Furthermore, within each story there are two protagonists, namely the chemists themselves and the molecules investigated in their chemical research efforts. Both can represent an excellent support to engage students beyond the curriculum content.

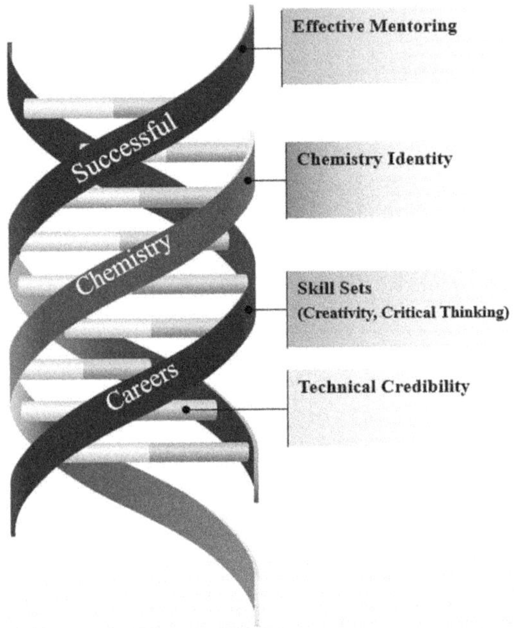

Figure 1. Emerging themes from the chapter narratives.

The beginning of this book is focused on the careers of chemists who were all raised in the segregated South of the United States, but persisted and earned doctoral degrees in chemistry. Both Dr. Bettye Washington-Greene and Dr. Nodie Monroe Washington had significant and successful careers working in industry focusing on the chemistry of latex and bleach. In addition, they further balanced their careers with their roles as wives and mothers—which is tremendously difficult. Dr. Washington-Greene's daughter shared that she doesn't know how her mother successfully balanced a career and family. Both Dr. Washington-Greene and Dr. Washington published their first papers in high-impact journals in the 1960s and registered multiple patents during their careers. Regarding the importance of gaining technical credibility in an organization Dr. Washington writes, "In order to gain technical credibility, you should have confidence in your abilities and have a positive mindset that you are technically sharp." She further emphasizes that gaining technical credibility means that your work must be grounded in "basic science and engineering principles." Technical credibility however isn't sufficient, effective mentoring is critical to gaining technical credibility within an organization. The publications and registered patents of these women clearly demonstrates they had significant technical credibility within their respective organizations. Furthermore, both latex and sodium nonanoyloxybenzensulfonate (NOBS) are important protagonists in their stories, which were critical to Washington-Greene and Washington gaining technical credibility and success.

Despite those achievements, Dr. Washington-Greene's daughter shared that toward the end of her career her mother dealt with frustrations as she saw colleagues with less experience—who she had trained—being promoted ahead of her. Dr. Nodie Washington shared that she took a serious risk early in her career challenging leadership to recognize and reward the achievements of Black women within the company. These women are chemistry trailblazers.

The chapter authored by Dr. Richard Evans serves two purposes, specifically telling his own chemistry journey and the story of his mentor, Professor Saint Elmo Brady. His chapter provides a student-centered perspective on the significance and impact of Professor Brady's career. A key point from Dr. Evans' chapter is the importance of effective mentoring, which leads to establishing 'chemistry identity.' Although not mentioned in the chapter, important protagonists in Professor Brady's journey are carboxylic acids, which are important organic molecules and can be explored at various points throughout the chemistry curriculum. He investigated the acidity of these organic molecules as a graduate student at the University of Illinois at Urbana-Champaign. Although Professor Brady did not publish his doctoral research in a peer-reviewed journal, he published three abstracts in *Science*. Furthermore, in 1939 Brady, while on the faculty at Fisk University, an HBCU (Historically Black Colleges and Universities) was the sole author of a paper published in the *Journal of the American Chemical Society* (*5*), a significant achievement that demonstrates his technical credibility.

The modern chemistry trailblazers provide shining examples of social entrepreneurship within the chemistry discipline. Social entrepreneurship essentially focuses on the establishment of a business to develop innovative solutions to address social justice issues (*6*). Dr. Darryl Boyd, a 2018 recipient of C&EN's Talented 12 Award (*7*) utilizes his chemistry training for public engagement through a series of online videos distributed through his company, Science Made Simple, LLC. Boyd celebrates women in his narrative stating, "I know the strength of women—especially Black and Latina women—and how driven, capable, and successful they are in the field of chemistry."

This self-proclaimed 'stinking chemist' is a doting father of a young daughter and describes the joys of fatherhood and highlights in his chapter the significance of women mentors during his career. It is tremendously important to have male voices celebrate women scientists. The protagonists featured in Boyd's story are sulfur-based polymers, with applications to fiber optics. Boyd has published his research efforts in high-impact journals including *ACS Macro Letters* (*8*). His research has garnered well over 600 citations, which demonstrates technical credibility.

The second social entrepreneur celebrated in the book is Dr. Pamela Leggett-Robinson, an organic chemist who established her own consulting company, PLR Consulting to specifically address pipeline and STEM education initiatives. This 'STEM Butterfly,' to borrow her own nickname, has focused her attention on providing sustainable support systems for women of color in STEM navigating academia (*9*). Her research is providing a critical narrative for women of color to be successful in the professoriate. An important protagonist in her story is the molecule dimethyldioxirane, an important oxidizing agent for chemical reactions.

Dagmawi Abebe, a physicist and filmmaker brought the story of Alice Augusta Ball to life in a narrative short film, "The Ball Method." Abebe emphasizes that Ball's story is inspirational because she was "defying gender and racial barriers," while simultaneously developing the first viable treatment for leprosy in the early 20th century. The protagonist in the story of Ball are ethyl esters, which she developed in the laboratory for treating leprosy. The stories of Boyd, Leggett-Robinson and Abebe emphasize the significance of the entrepreneurial mindset for a successful career pathway in STEM.

The remaining four modern trailblazers celebrated in this book are leading the way with innovations in teaching, research and academic administration. Professor Angela W. Peters, a biochemist and Professor Leyte Winfield, an organic chemist, are doing amazing work on the campuses of two HBCUs in the State of Georgia: Albany State University and Spelman College, respectively. Effective mentoring is critical to the success of these women in academia. Both Peters and Winfield have served as department chairs at their respective institutions. Prior to her appointment as Provost and Vice President of Academic Affairs at Albany State University, Professor Peters served as Vice Provost for Academic Programs at Claflin University, an HBCU in Orangeburg, South Carolina. Throughout her career she has focused on innovation in both teaching (*10*) and professional development of teachers and educators (*11*). The protagonist in Professor Peters' chemistry journey is human thymidylate synthase, which is an important target for cancer therapeutics.

HBCUs are important institutions for training the next generation of African American scientists (*12*). Professor Winfield recently stated regarding their significance, "We're training the whole student, and we give them a cultural experience that connects with how they identify (*12*)." A graduate of Dillard University, an HBCU in New Orleans and the University of New Orleans, Winfield's research is focused on the anti-cancer behavior of organic molecules, which are key protagonists in her story (*13*). Professor Winfield focuses on innovation in teaching organic chemistry and diversity, equity and inclusion in the chemical sciences (*14*, *15*), and has published her studies in high-impact journals.

Professor Rená Robinson, an analytical chemist teaching at Vanderbilt University and a 2016 recipient of C&EN's Talented 12 Award (*16*) is truly passionate about understanding Alzheimer's disease as well as health disparities. She decided to pursue a career in academia to serve as a role model and mentor to women and chemists of color. Robinson is a prolific researcher with publications in high-impact journals such as *Nature*, *Journal of Proteomics*, and the *Journal of the American Chemical Society*. Robinson's articles have attracted well over 2,000 citations, which is significant evidence of her technical credibility.

LaVetta Appleby, a chemistry and education professor teaching at Lawrence Technological University in Michigan shared in her narrative that her "journey to become a chemist was by no means a smooth one." This is an important theme that connects all of the chemists highlighted in this book. There are several protagonists in her story including ammonia, phosphine and the fictional vibranium. Professor Appleby has done tremendous work with teachers and educators delivering innovative professional development workshops focused on forensic science to improve classroom teaching and student engagement. In the past, chemistry education has fallen short of including a diversity of chemists—innovation in teaching is absolutely critical to remedy this.

Next Steps: Recommendations for Inclusion in Chemistry

What are some important steps for chemistry professors to address equity in their classrooms? How can educators make key connections with their students in the classroom using culturally relevant pedagogy? One tangible option is to extend an invitation to chemists profiled in this book—or others with similarly inspiring stories—to deliver a seminar during the academic year in your department. Given the impact of the pandemic caused by the coronavirus (COVID-19) on higher education and many institutions pivoting to online teaching, research seminars can be delivered remotely as well, removing some of the barriers that might have been present previously—for example related to traveling. In addition, chemistry faculty can have their students critically review a journal article published by the chemists from historically marginalized groups. A chemistry literature reading assignment is an excellent approach to bringing research into the classroom, which can be aligned with course content. Class discussions could include a 'surprise' Zoom session with the authors, which would certainly excite your students. This type of engagement could also potentially provide opportunities for research collaborators across institutions.

Another recommendation for chemistry professors is to incorporate the lesson plans included in the appendices of the chapters of this book to enhance your classroom instruction. These are tangible teaching strategies to address equity in the classroom. Finally, in order to make cultural connections with your students in the classroom try watching television shows and popular movies through the lens of a chemist. This approach will help you to recognize and identify teachable moments that connect to chemistry content, which can be used as a "hook" for engagement with your students. Thus, you are not changing the chemistry content, only how it is delivered to your students.

Chapter Summary

African American Chemists: Academia, Industry, and Social Entrepreneurship celebrates the achievements of chemists, who have gained technical credibility within their organizations through innovative teaching and research, high-impact publications and successful career pathways in chemistry. Their career pathways in the chemical sciences emphasizes the significance of grit, persistence and perseverance. It is absolutely imperative that we all work diligently and collectively to broaden the image of a chemist in our society, which can be achieved through recommendations discussed previously. Thus, storytelling in the classroom is a powerful tool to engage the next generation of chemistry leaders (*17*). Broadening the image of a chemist is critically important for us to effectively address diversity, equity and inclusion within our discipline.

References

1. Rifkin, M. Addressing Underrepresentation: Physics Teaching for All. *The Physics Teacher*. https://aapt.scitation.org/doi/10.1119/1.4940167 (accessed Nov 8, 2020).
2. Guest Post: Teaching Social Justice in the Physics Classroom, Part 1. *Quantum Progress*. https://quantumprogress.wordpress.com/2015/02/12/guest-post-teaching-social-justice-in-the-physics-classroom-part-1/ (accessed Nov 8, 2020).
3. Schinske, J. N.; Perkins, H.; Snyder, A.; Wyer, M. Scientist Spotlight Homework Assignments Shifts Students' Stereotypes of Scientists and Enhance Science Identity in a Diverse Introductory Class. *CBE–Life Sciences Education*. https://www.lifescied.org/doi/10.1187/cbe.16-01-0002 (accessed November 8, 2020).
4. Jarvis, C. L. Helping Students Re(think) of Themselves as Scientists. *Chemical and Engineering News*. https://cen.acs.org/education/undergraduate-education/Helping-students-rethink-themselves-scientists/98/i28 (accessed Nov 8, 2020).
5. Brady, S. E. Ricinus communis. I. Oxidation of Ricinoleic Acid. *Journal of the Am. Chem. Soc.* **1939**, *61*, 3464–3467.
6. Peek, S. What is Social Entrepreneurship? 5 Examples of Businesses with a Purpose.. https://www.uschamber.com/co/start/startup/what-is-social-entrepreneurship (accessed Feb 1, 2021).

7. Torrice, M. Darryl Boyd: Polymers Pioneer is Stitching Sulfur into Molecular Strands to Create Advanced Optics. *Chemical and Engineer News.* https://cen.acs.org/materials/polymers/Darryl-Boyd/96/i33 (accessed Nov 11, 2020).
8. Boyd, D. A.; Nguyen, V. Q.; McClain, C. C.; Kung, F. H.; Baker, C. C.; Myers, J. D.; Hunt, M. P.; Kim, W.; Sanghera, J. S. Optical Properties of a Sulfur-Rich Organically modified Chalcogenide Polymer Synthesized via Inverse Vulcanization and Containing an Organometallic Comonomer. *ACS Macro Lett.* **2019**, *8* (2), 113–116.
9. Robinson-Leggett, P. M.; Villa, B. *Overcoming Barriers for Women of Color in STEM: Emerging Research and Opportunities*; IGI Global: Hershey, PA, 2020.
10. Peters, A. W. Teaching Biochemistry at a Minority-Serving Institution: An Evaluation of the Role of Collaborative Learning As a Tool for Science Mastery. *Journal of Chemical Education* **2005**, *82* (4), 571.
11. Li, N.; Peters, A. W. Preparing K-12 Teachers for ELLS: Improving Teachers' L2 Knowledge and Strategies through Innovative Professional Development. *Urban Education* **2016**, 55 (10), 1489–1506.
12. Widener, A. Who Has the Most Success Preparing Black Students for Careers in Science?" Historically Black Colleges and Universities. *Chemical and Engineering News.* https://cen.acs.org/education/success-preparing-Black-students-careers/98/i34 (accessed Nov 11, 2020).
13. Winfield, L. L. *Synthesis and Anti-Proliferative Effect of Benzimidazole Derivatives.* U.S. Patent 7,947,723, 2011.
14. Leontyev, A.; Houseknecht, J. B.; Maloney, V.; Muzyka, J. L.; Ross, R.; Welder, C. O.; Winfield, L. OrganicERs: Building a Community of Practice for Organic Chemistry Instructors Through Workshops and Web-Based Resources. *Journal of Chemical Education* **2020**, 97 (1), 106–111.
15. Wilson-Kennedy, Z. S.; Payton-Stewart, F.; Winfield, L. L. Toward Intentional Diversity, Equity and Respect in Chemistry Research and Practice. *Journal of Chemical Education* **2020**, 97 (8), 2041–2044.
16. Robinson, R. *2016 C&EN's Talented 12 Chemical and Engineering News.* http://talented12.cenmag.org/rena-robinson/ (accessed Nov 11, 2020).
17. Collins, S. N. The Importance of Storytelling in Chemical Education. *Nat. Chem.* **2020**, *13*, 1–2. https://doi.org/10.1038/s41557-020-00617-7 (accessed Jan 6, 2021).

Editor's Biography

Sibrina N. Collins

Sibrina N. Collins earned a Ph.D. in inorganic chemistry from The Ohio State University and is the Founding Executive Director of the Marburger STEM Center on the campus of Lawrence Technological University. In 2018, she received the Real McCoy Award in Science from the Detroit Area Pre-College Engineering Program (DAPCEP) for her leadership and public engagement efforts in STEM (Science, Technology, Engineering and Mathematics) education. Her research efforts include inorganic chemistry, chemical education, history of chemistry, and addressing inclusion and equity in the STEM fields. Dr. Collins is an active member of the American Chemical Society (ACS), serving as an Associate Member on the National Historic Chemical Landmark (NHCL) Subcommittee. She has published in ACS journals including *Inorganic Chemistry*, the *Bulletin for the History of Chemistry* and the *Journal of Chemical Education.* She has authored a book chapter in the ACS Symposium Series Volume 1370 entitled "Inorganic Chemistry: Vibranium and Marvel Studios' *Black Panther*," and "Critical Mass Takes Courage: Diversity in the Chemical Sciences" published in the ACS Symposium Series Volume 1256. In 2020, Dr. Collins' career pathway was featured in the "Career Ladder" spotlight published by *Chemical and Engineering News*.

Indexes

Author Index

Subject Index

Printed in the USA/Agawam, MA
January 21, 2022

788130.002